Bismuth-Ethanedithiol:
A Potential Drug to Treat Biofilm Infections of Medical Devices Produced by *Staphylococcus Epidermidis* and *Proteus Mirabilis*

Jithendra Gunawardana

DISSERTATION.COM

Boca Raton

Bismuth-Ethanedithiol:A Potential Drug to Treat Biofilm Infections of Medical Devices Produced by Staphylococcus Epidermidis and Proteus Mirabilis

Dissertation.com
Boca Raton, Florida
USA • 2010

ISBN-10: 1-59942-304-9
ISBN-13: 978-1-59942-304-3

ABSTRACT

Staphylococcus epidermidis is the leading cause of hospital-acquired infections associated with implanted medical devices. Likewise, *Proteus mirabilis* is one of the leading causes of nosocomial Urinary Tract Infections (UTIs) and is associated with urinary catheter blockage. Both organisms produce biofilms, which make them less susceptible to antibiotics. Therefore, these infections are often notoriously difficult to treat and in many cases lead to complications. Preventing biofilm formation or killing a pre-existing biofilm would be useful solutions since biofilm formation is a phenomenon commonly observed in these organsims.

This study investigated the *in vitro* efficacy of Bismuth-ethanedithiol (BisEDT) against biofilms produced by *P. mirabilis* and *S. epidermidis*. Anti-biofilm activity of BisEDT was evaluated in terms of inhibiting/preventing biofilm formation and eradicating/killing pre-formed biofilm produced by single species *S. epidermidis* and *P. mirabilis* as well as dual species biofilms produced by these two organisms. The following assays were used to assess: 1) Polystyrene bead assay. 2) Agar diffusion susceptibility assay. 3) Spectrophotometric bead assay. The assays showed that 0.1 µg/ml of BisEDT inhibited *S. epidermidis* biofilm formation by 99.9% in single species biofilms. One µg/ml of BisEDT inhibited single species *P. mirabilis* biofilm formation by 99.9%. Dual species biofilms of both organisms were inhibited by 99.9% by five days of exposure to 0.5 µg/ml. 99.9% biofilm eradication in single and dual species biofilms of both organisms was achieved by 5 µg/ml of BisEDT.

These results suggest that BisEDT was highly successful in inhibiting and eradicating biofilms. Therefore, this drug may be of use to treat device related infections caused by *S. epidermidis* and *P. mirabilis*.

ACKNOWLEDGEMENTS

I would like to take this opportunity to acknowledge all the people who helped me complete my Master's Thesis work. First, I would like to thank my major professor, Dr. Jay F. Sperry for his fine direction and words of encouragement throughout, as well as providing all necessary tools to complete this project. I am also thankful to my committee members Drs. D. Laux, H. Bibb and R. Rhodes Warm thanks also go out to other faculty members, staff and graduate students at CMB.

I am especially grateful for my family for their constant encouragement and immense support that enabled me to pursue my academic goals. Without your love and blessings, this would not have been possible. My gratitude cannot be expressed in words.

Much appreciation is extended to Dr. Philip Domenico (Winthrop-University Hospital, NY) for providing us with valuable advice and the bismuth-ethanedithiol used in this project.

PREFACE

The following thesis is written in the manuscript format and conforms to guidelines established by the Graduate School of the University of Rhode Island and the *ASM Style Manual for Journals and Books* (American Society for Microbiology, 2000.)

TABLE OF CONTENTS

LIST OF TABLES

LIST OF FIGURES

INTRODUCTION

A recent announcement from the National Institutes of Health reported that "more than 60% of all microbial infections are caused by biofilms" (19). This percentage represents biofilm-associated medical conditions such as dental plaque, gingivitis, urinary tract infections, middle ear infections as well as bacterial colonization of indwelling medical devices. The last is a major medical concern due to significant rates of patient morbidity/mortality and high economic cost (25). This is partly due to modern medicine's increased dependency on indwelling devices such as intravascular catheters, cerebrospinal fluid shunts, continuous ambulatory peritoneal dialysis catheters, orthopedic devices, artificial joints, prosthetic heart valves and cardiac pacemakers as tools of diagnosis and/or therapy for various medical conditions (17). Colonization of these devices contribute to nosocomial (hospital-acquired) infections, particularly endocarditis and blood-stream infections. A recent survey revealed that up to 2 million patients a year are affected by nosocomial infections, causing 88,000 deaths and costing over $4.5 billion to the health care system (26).

Coagulase-negative staphylococci (CNS), especially *Staphylococcus epidermidis* is the most common cause of infections from both temporarily and permanently inserted medical devices (17, 21). For instance, this organism is the culprit of up to 67% of central nervous system shunt complications, 70% of catheter-related infections, 50% of prosthetic cardiac valve infections and 50% of joint replacement infections (22). Treatment of these infections is further complicated by the emergence of antibiotic resistance. With the onset of methicillin-resistant CNS, vancomycin became the drug of hope, which also in recent years has exhibited decreased sensitivity to CNS (29, 33).

Proteus mirabilis is the third most common pathogen involved in Urinary Tract Infections (UTIs; after *Escherichia coli* and *Klebsiella* spp.) (4). UTIs make up almost 40% of nosocomial infections of which, as much as 80% are associated with indwelling urinary catheters. The risk of death, once catheter-related bateriuria occurs increases by 2.8 fold and additional cost of hospitalization is increased by $2800 (2, 27). Urinary catheters colonized by *P. mirabilis* produce encrustations (due to urease activity), which not only blocks the catheter but may also cause bladder stones (2, 31). Treatment of *P. mirabilis* infections is also affected by antibiotic resistance. Stickler et al. (30) showed that this organism survived well in the presence in chlorhexidine. Likewise *P. mirabilis* has been shown to develop resistance after 14 days of exposure to mandelic acid, a drug widely used to treat UTIs (2).

Both nosocomial *S. epidermidis* and *P. mirabilis* involved in device-related infections produce biofilms. Two leading scientists in the field have provided classic definitions of biofilms; Costerton et al. defines biofilms as "a structured community of bacterial cells enclosed in a self-produced polymeric matrix and adherent to an inert or living surface" (7). Elder et al. defines biofilms as "a functional consortium of microorganisms organized within an extensive exopolymer matrix" (16). Apart from bacterial cells and the exopolymer matrix (also known as Extracellular Polymeric Substance (EPS), glycocalyx or more commonly 'slime', which is primarily composed of polysaccharides and proteins), a biofilm may contain other materials depending on the environment in which it is found. For example, inflammatory response proteins, complement, fibrinogen, fibronectin and glycosaminoglycans may be found in biofilms associated with artificial prosthetic devices, while mineral crystals, clay, silt and corrosion particles may be found in industrial or portable water systems and natural aquatic systems (15, 16).

The process of biofilm formation and its metabolism is highly complex. The initial steps involved in bacterial adherence and biofilm production on a surface depends on several factors such as the texture of the surface, chemical alteration of the surface due to materials found in the medium (called 'conditioning'), hydrodynamics, pH, temperature, presence of ions and nutrients in the medium (15). More important are the properties of the bacterial cell such as the presence of flagella and/or pili and the ability to produce EPS. Biofilm production is a cell-density dependent process where quorum-sensing systems sense the number of planktonic cells in close proximity to a conditioned surface and begin to adhere to it as a result of hydrophobic and electrostatic interactions (16, 23). Investigators have shown that a gene cluster (called the *ica* operon in staphylococci species) encodes the production of polysaccharide intercellular adhesion (PIA), which mediates cell aggregation once attachment has occurred (8, 16). Next, the superficially attached cells secrete EPS, which acts as a glue that adheres them to the foreign surface more tightly and provides protection to the cells within the biofilm. The final step in biofilm formation involves the maturation of the multi-layers characterized by the replication of attached organisms (16).

Bacterial cells enclosed within biofilms are resistant to antimicrobial agents and the host immune system than planktonic cells. Researchers have suggested several mechanisms as to how biofilms successfully evade antimicrobial agents. One mechanism is the restricted penetration of antimicrobial agents due to the multi-layer structure of biofilms. Investigators have shown that these layers act as a diffusion barrier or a molecular filter, physically retarding and/or preventing the penetration of antibiotics and antimicrobial proteins (i.e. lysozyme) and complement (19). Further evidence shows that, in some cases (depending on the species and the agent used) the antimicrobial agent is rendered inactive before it can diffuse through the layers of the biofilm, possibly due to binding by EPS (7, 19). Another important observation was made during experiments involving resistance of *Pseudomonas aeruginosa* biofilms to tobramycin and *S. epidermidis* slime to vancomycin and teichoplanin. In both organisms, researchers found the minimum inhibitory concentration (MIC) required to sterilize biofilms were 5 and 15 fold more than their planktonic counterparts, respectively (16, 19). A second mechanism relates to the growth rate of cells within biofilms. Cells that are buried deep in the layers are metabolically inert (possibly due to nutrient limitations and waste build-up). Therefore, these cells are not actively growing. The function of these cells is primarily to provide a foothold on the surface of the foreign material (to act as a 'suction cup'), so that multi-layers may be formed on them. Most antibiotics target actively dividing cells. For example, cell growth is an absolute requirement for penicillin and ampicillin to be effective. Also, the rate of kill of the target cell is proportional to the rate of growth of the cell (19). Other speculative reasons that would explain the resistance of biofilms to treatment are lower pH and higher CO_2 within these slime layers, expression of possible resistance genes, changes in cell wall composition and phenotypic variation in enzymatic activity (16, 19, 25).

Since bacterial biofilms show poor response to antibiotics, they are notoriously difficult to treat without the removal of the infected device. As a result, several remedies are currently being developed in order to eradicate/inhibit colonization of medical devices. The use of electromagnetic fields, ultrasound, genetic manipulation of 'biofilm genes' and coating or impregnating implants with antimicrobial agents are several options that are being investigated (19, 22, 25). This study proposes a relatively new antimicrobial agent that can be effectively used against catheter-related infections.

Bismuth-ethanedithiol (BisEDT) is a new class of anti-biofilm agents that has demonstrated remarkable activity against a broad spectrum of biofilm-producing organisms. Several investigators using Bismuth-dimercaprol (BisBAL), Bismuth-ethanedithiol and Bismuth-3,4-dimercaptotoluoene (BisTOL) have shown inhibition of *Klebsiella* spp. capsule production, reduction of alginate expression by *P. aeruginosa* and inhibition of slime-producing *S. epidermidis* (10, 14, 18, 35). Bismuththiols, while having up to 1000-fold greater antimicrobial activity than other bismuth salts, inhibit slime expression in both gram positive and negative organisms at non-toxic concentrations (10).

To date, the effects of BisEDT on *S. epidermidis* slime production has not been investigated by a polystyrene bead adherence assay. The effects of BisEDT on *P. mirabilis* biofilms have not been reported. Furthermore, treatment of a mixed biofilm of both organisms with BisEDT has not been attempted. Therefore, this study investigates the *in vitro* efficacy of BisEDT in:

> **Study #1:** Inhibiting/preventing biofilm production by *S. epidermidis*, *P. mirabilis* and a mixture of both organisms.

> **Study #2:** Eradicating/killing pre-formed biofilm produced by the above organisms.

MATERIALS AND METHODS

Bacterial strains: Both *S. epidermidis* RP62A (ATCC 35984) and *P. mirabilis* (ATCC 12453) were obtained from the American Type Culture Collection (ATCC.)

Preparation of BisEDT (1:1): Both liquid and powder forms were prepared. For the liquid form, 4.85 mg of bismuth nitrate (BN) was dissolved in 1 ml of 1,2-propanediol, while 100 μl of 1,2-ethanedithiol (EDT) was dissolved in 900 μl of 1,2-propanediol. 1000 μg/ml working stock BisEDT was prepared by adding 50 μl of BN, 5.8 μl of concentrated HCl (12.1N) and 4.2 μl of EDT to 940 μl of 1,2- propanediol. For the non-liquid form, 1 mg of BisEDT powder was placed in 990 μl of 1,2 propanediol and 10 μl of concentrated HCl. For both types, a water bath sonicator (Fischer FS 3 ultrasonic cleaner, Pittsburgh, PA) was employed to break up particles. (Five micromoles of BT is approximately equal to 1 μg of bismuth/ml.) The working stock was diluted

1:10 and 1:100 in sterile saline. BN was made fresh bimonthly while EDT was made biweekly.

Preparation of initial inoculum: Both bacterial strains were grown overnight, shaking at 150 rpm (Model G-10, New Brunswick Scientific Co., Inc., Edison, NJ) at 37°C in trypticase soy broth (TSB; Difco, Detroit, MI.) The OD_{600} of the overnight cultures was adjusted to 0.01 (~10^7 CFU/ml) and 15 µl aliquots were used as the inocula for each culture tube. RP62A or *P. mirabilis* inoculum was used for single-species biofilm, while both were used to establish a dual-species biofilm.

Statistical analysis: All tests were done in duplicate in three different trials. The statistical significance of the data was determined by a paired t-test using Slide Write Plus software (Version 4.0 by Advanced Graphics Software, Inc., Carlsbad, CA.) A *P* value of < 0.05 was considered to be statistically significant. The ± standard error of the mean (SEM) was also recorded.

MIC determinations of BisEDT: The minimum inhibitory concentration of BisEDT against planktonic cells of both organisms was determined by a broth microdilution method as recommended by the National Committee for Clinical Laboratory Standards (NCCLS; 34). The broth medium used was cation-adjusted Mueller Hinton broth (CAMHB; Difco.)

Study #1: <u>Inhibition of biofilm formation</u>

Method #1: Polystyrene bead assay

Adherence assay: Each culture tube contained 15 µl inoculum, one sterile polystyrene bead (5.5mm diam., Precision Plastic Ball Co., Franklin Park, Illinois) and 150 µl of CAMHB. The following concentrations (final) of BT were added to the culture tubes:

Tube	BisEDT (final conc. µg/ml)
Reference	0
#1	0.1 (only for single-species RP62A)
#2	0.5
#3	1
#4	5

The final volume of all tubes was brought up to 300 µl with sterile de-ionized water, mixed well and incubated at 37°C with shaking at 150 rpm for 1, 3 and 5 days.

Following incubation, the liquid from all tubes was aspirated, the beads were washed 3 times with sterile phosphate-buffered saline (PBS, pH 7) and 500 µl of a solution containing 0.5% Tween 80 (polyxyethylene-sorbitan monooleate; Sigma, St. Louis, MO) and 10mM EDTA (ethylenediamine tetraacetic acid; Sigma) was added to each tube. After 10 minutes of Tween 80-EDTA treatment, bacteria adhered to the beads were released by vigorous vortexing (Fischer Vortex-Genie 2™ Model G-560, Scientific Industries, Inc., Bohemia, NY) for 3 minutes in the Tween 80-EDTA solution. An ultrasonic bath was not used to release adhered cells since previous studies indicate the vortex only method to have a recovery efficiency of 97% (24). The liquid was serially diluted and the number of biofilm producing cells that adhered to the beads were spot plated and enumerated by the viable plate count method on selective media. Mannitol salts agar (MSA; Difco) plates were used to select for the growth of *S. epidermidis* while inhibiting the growth of Gram negative, *P. mirabilis*. Xylose lysine deoxycholate (XLD; Difco) plates were used to select for *P. mirabilis* and to inhibit Gram positive, *S. epidermidis* growth. Preliminary data (not shown) showed insignificant differences between the number of colonies grown on selective and non-selective, i.e. blood agar media (Difco.) All plates were incubated overnight at 37˚C and the colonies counted to determine CFU/ml.

Determination of treatment effects: Treatment effects were determined by the relative inhibition of biofilm production (expressed as a mean percentage):

100 - [(CFU of treated bead ÷ CFU of reference bead) ×100]

The CFUs generated by the untreated/reference bead served as the reference inoculum for that set of experiments.

Method #2: Agar diffusion susceptibility test

Disk preparation: Mueller-Hinton agar (Difco) plates were streaked for confluent growth with RP62A or *P. mirabilis* for single-species biofilm. Sterile filter paper disks (6 mm diameter) soaked with10 µl of 0, 0.1, 0.5, 1, 5 and 10 µg BisEDT were placed on the surface of agar plates and incubated overnight at 37˚C.

Determination of treatment effects: After incubation, the diameter of the zone of inhibition created by the diffusion of the drug was measured in millimeters (mm) using a scientific ruler (Sorvall, Norwalk, CT.)

Method #3: Spectrophotometric assay

Adherence assay: Culture tubes with polystyrene beads were inoculated, the same concentrations of BisEDT were added and incubated under conditions described in method #1. Following incubation, non-adherent cells were removed by rinsing with PBS. Biofilm on the beads was stained with 350 µl of 1% crystal violet (Fisher) for 25 min. followed by further rinsing with distilled water. The cell-associated stain was solubilized in 400 µl of dimethyl sulfoxide (DMSO; Fisher) for 5 min. as described in (3).

Determination of treatment effects: The resulting solution was quantified by measuring the OD at 570 nm using an UV spectrophotometer (Pharmacia-Biotech.) An uninoculated/untreated bead stained with crystal violet and subsequently solubilized in DMSO was used as the blank.

Study #2: Eradication of pre-formed biofilm

Method #1: Polystyrene bead assay

Adherence assay: 15 µl aliquots of the inocula were added to culture tubes containing a sterile polystyrene bead and 150 µl of CAMHB followed by overnight incubation at 37°C with shaking at 150 rpm. The next day the following concentrations (final) of BisEDT were added to the tubes containing the pre-formed biofilm:

Tube	BisEDT (final conc. μg/ml)
Reference	0
#1	5
#2	10

The final volume of all tubes was brought up to 300 μl with sterile de-ionized water, mixed well and incubated at 37°C with shaking at 150 rpm for 1, 3 and 5 days. The removal of adherent bacterial cells and enumeration was done as previously described.

Determination of treatment effects: Treatment effects were determined by the relative eradication/killing of pre-formed biofilm (expressed as a mean percentage):

$$100 - [(\text{CFU of treated bead} \div \text{CFU of reference bead}) \times 100]$$

The CFUs generated by the untreated/reference bead served as the reference inoculum for that set of experiments.

Method #2: Spectrophotometric assay

Adherence assay: Culture tubes with polystyrene beads were inoculated and incubated under conditions described in method #1. 1, 5 and 10 μg/ml final concentrations of BisEDT were added to the existing biofilm-coated beads. Following incubation, removal of non-adherent cells, biofilm staining and solubilizing was done as described in study #1.

Determination of treatment effects: The resulting solution was quantified by measuring the OD at 570 nm using an UV spectrophotometer. An uninoculated/untreated bead was used as the blank.

Examination of bacterial antagonism: 15 µl aliquots of RP62A and *P. mirabilis* were used to inoculate polystyrene beads and incubated overnight as previously described, but without the addition of BisEDT. Following incubation, the adherent cells were released, serially diluted and plated on corresponding media as described previously. Bacterial antagonism exhibited by one organism on the other was calculated as a mean percentage:

$$100 - [(\text{CFU of bead inoculated} \div \text{CFU of bead inoculated}) \times 100]$$
with both organisms   with one organisms

RESULTS

MIC determinations. The MIC of BisEDT against planktonic *S. epidermidis* RP62A was 0.1 µg/ml and 0.5 µg/ml against planktonic *P. mirabilis* (Table 1 and Table 2).

<u>Inhibition/Prevention of biofilm formation</u>

Polystyrene bead assay. Both 0.1 and 0.5 µg/ml of BisEDT inhibited *S. epidermidis* biofilm. Biofilm inhibition by 0.1 µg/ml was 90.6%(±5.6), 69%(±13.9) and 99.6%(±0.3) at 1, 3 and 5 days respectively. At 0.5 µg/ml, 97.1%(±2.3), 97.5%(±2.2) and 99.9%(±0.01) inhibition was seen at 1, 3 and 5 days, respectively (Table 3, Figure 1). However, 0.5 µg/ml was not statistically different than 0.1 μg/ml ($P > 0.05$). *Staphylococcus epidermidis* biofilm inhibition when coexisting as part of a dual species biofilm with *P. mirabilis* was 65.1%(±3.5), 48.6%(±11) and 85.4%(±7.5) at 0.5 μg/ml of BisEDT at 1, 3 and 5 days of exposure. Inhibition at 1 μg/ml was 75.6%(±6.7), 93.4%(±6.5) and 99.9%(±0.06) at 1, 3 and 5 days, respectively. At 5 μg/ml inhibition was 99.5%(±0.3), 99.6%(±0.6) and 99.9%(±0.005) at 1, 3 and 5 days (Table 4, Figure 2). At day 1, 5 μg/ml was statistically more effective than 0.5 or 1 μg/ml ($P < 0.05$). At day 3, 5 μg/ml was significantly better than 0.5 μg/ml ($P < 0.05$), but not 1 μg/ml ($P > 0.05$). By day 5, both 0.5 and 1 μg/ml were equally as effective as 5 μg/ml ($P > 0.05$).

Single species *P. mirabilis* was inhibited 21%(±9.3), 52.4 %(±5) and 54.6%(±11.8) at 1, 3 and 5 days, respectively at 0.5 μg/ml. At 1 μg/ml the inhibition of biofilm formation was 99.7%(±0.1), 60.9%(±12.2) and 75.8%(± 9.4). At 5 μg/ml, inhibition remained at 99.9% through out the 5 days (±0.8, 0.3 and 0.04 for 1, 3 and 5 days, respectively; Table 5, Figure 3). At day 1, 5 μg/ml was significantly better than 0.5 μg/ml ($P < 0.05$), but not 1 μg/ml ($P > 0.05$). At day 3, 5 μg/ml treatment was most effective ($P < 0.05$). At day 5, 5 μg/ml was statistically better than 0.5 μg/ml ($P < 0.05$), but no significantly different than 1 μg/ml ($P > 0.05$). This bacteria in a mixed biofilm with *S. epidermidis* was inhibited by 50.5%(±4.5), 47.9% (±6.7) and 81.9%(±10.1) at 1, 3 and 5 days at 0.5 μg/ml. At 1 μg/ml, 90.8%(±1.8), 77.3%(±7.7) and 99.2%(±0.6) of inhibition was observed. At 5 μg/ml, 95.9%(±1.7), 98.8%(±0.8) and 99.9%(±0) inhibition was seen over the same time period

13

(Table 6, Figure 4). At days 1 and 3, 5 µg/ml treatment proved to be the most effective ($P < 0.05$). However, at day 5, 5 µg/ml was not statistically better than the other lower concentrations ($P > 0.05$).

Agar disk diffusion assay. An increase in the zone of inhibition was seen with the increase in BisEDT concentration. *S. epidermidis* inhibition was 15 mm($\pm$0.4) at 0.1 µg, 17 mm($\pm$0.8) at 0.5 µg, 20 mm($\pm$0.2) at 1 µg, 21 mm ($\pm$0.3) at 5 µg and 24 mm($\pm$0.2) at 10 µg of BisEDT. No inhibition of *P. mirabilis* was seen at 0.1, 0.5 and 1 µg of BisEDT. 8 mm ($\pm$0.1) and 10 mm($\pm$0.2) inhibition zones were seen at 5 and 10 µg, respectively (Table 7, Figure 5).

Spectrophotometric bead assay. The absorbance at OD_{570} of the reference (untreated) *S. epidermidis* bead was 1.62($\pm$0.3). When treated with 0.1 µg/ml of BisEDT, the absorbance was 1.64($\pm$0.3). The absorbance was 1.62($\pm$0.2) at 0.5 µg/ml of treatment, 1.35($\pm$0.3) at 1 µg/ml and 1.26($\pm$0.2) at 5 µg/ml (Table 8, Figure 6).

For *P. mirabilis*, absorbance at OD_{570} of the reference/untreated bead was 1.67($\pm$0.2), 1.44($\pm$0.3) when treated with 0.1 µg/ml of BisEDT, 1.73($\pm$0.3) at 0.5 µg/ml, 1.26($\pm$0.2) at 1 µg/ml, 1.58($\pm$0.2) at 5 µg/ml and 1.1($\pm$0.2) at 10 µg/ml of the drug (Table 8, Figure 6).

Eradication/Killing of biofilm

Polystyrene bead assay. There was no statistical difference between BisEDT concentrations 5 µg/ml and 10 µg/ml for both *S. epidermidis* and *P. mirabilis* biofilm eradication. Biofilm eradication of single species *S. epidermidis* at 5 μg/ml was 95.4% (±1.8), 98.5%(±1.8) and 89.9%(±8.1) at 1, 3 and 5 days respectively. At 10 μg/ml 98.5%(±1.3), 99.7%(±0.1) and 96.6%(±3.2) of reduction in cell number was seen at 1, 3 and 5 days, respectively (Table 9, Figure 7). RP62A biofilm eradication when formed as part of a dual species biofilm with *P. mirabilis* was 98.4%(±1.1), 97.6%(±2) and 98.8%(±0.5) at 5 μg/ml of BisEDT at 1, 3 and 5 days of exposure. Eradication at 10 μg/ml was 99.7%(±0.1), 99.9%(±0.006) and 99.9%(±0.002) at 1, 3 and 5 days, respectively (Table 10, Figure 8). Treatment with10 μg/ml BisEDT was not significantly different than treatment with 5 μg/ml for either single species or dual species *S. epidermidis* throughout the 5 days of exposure ($P > 0.05$).

Single species *P. mirabilis* was killed 90.2%(±2.2), 94.4 %(±3.3) and 96.1%(±1.7) at 1, 3 and 5 days, respectively at 5 μg/ml. At 10 μg/ml biofilm eradication was 99.9%(±0), 99.9%(±0.008) and 99.8%(± 0.01) (Table 11, Figure 9.) At day 1, 10 μg/ml was more effective than 5 μg/ml ($P<0.05$), while at days 3 and 5, there were no significant differences between the concentrations ($P > 0.05$). Viable *P. mirabilis* cells in a mixed biofilm with *S. epidermidis* were reduced by 90%(±0.8), 99.7% (±0.1) and 98.8%(±0.4) at 1, 3 and 5 days at 5 μg/ml. At 10 μg/ml, 99.9%(±0.01), 99.9%(±0.008) and 99.9%(±0) of eradication was seen (Table 12, Figure 10). There was no significant difference between 5 and 10 μg/ml through out the 5 days ($P > 0.05$).

Spectrophotometric bead assay. The absorbance at OD_{570} of the reference *S. epidermidis* bead was 1.52($\pm$0.2). At 1, 5 and 10 μg/ml of BisEDT treatment, the values are 1.31($\pm$0.7), 1.21($\pm$0.6) and 0.72($\pm$0.1) respectively (Table 13, Figure 11). The reference *P. mirabilis* bead was 1.85($\pm$0.4) when measured at the same wavelength. At 1, 5 and 10 μg/ml of BisEDT treatment, the values were 1.62($\pm$0.3), 1.56($\pm$0.3) and 1.35($\pm$0.3) respectively (Table 13, Figure 11).

Bacterial antagonism

Polystyrene bead assay. No significant levels of antagonism between *S. epidermidis* and *P. mirabilis* were observed ($P < 0.05$). Single species *S. epidermidis* growth was $log_{10}CFU$ 8.57($\pm$0.6). When grown in the presence of *P. mirabilis*, it increases insignificantly to $log_{10}CFU$ 8.67($\pm$0.2). Single species *P. mirabilis* growth was $log_{10}CFU$ 8.8($\pm$0.1). When grown in the presence of *S. epidermidis,* it decreases insignificantly to $log_{10}CFU$ 8.77($\pm$0.1). (Table 14, Figure 12)

DISCUSSION

This study was done to investigate the *in vitro* efficacy of BisEDT against *S. epidermidis* RP62A (ATCC 35984) and *P. mirabilis* (ATCC 12453) biofilms. The anti-biofilm effectiveness of this drug was evaluated against both single species and dual species biofilms. These bacterial strains were chosen for the study due to their remarkable ability to produce biofilms (5, 30).

The results of study #1 demonstrate that BisEDT successfully inhibited biofilm formation of both organisms. This was demonstrated by the polystyrene bead and the disk diffusion assays. BisEDT (0.1 μg/ml) effectively inhibited *S. epidermidis* biofilm formation on polystyrene beads over a period of 5 days. Five micrograms per ml was the most effective in inhibiting *S. epidermidis* in a mixed biofilm with *P. mirabilis* at day 1. However, the concentration required decreased to 1 μg/ml by day 3 and further to 0.5 μg/ml at day 5. Statistically, these lower drug concentrations prevented biofilm formation as effectively as 5 μg/ml. Therefore, indicating that a lower dose over a longer period of exposure is as effective as a higher dose.

Some fluctuation in drug dosage was observed for single species *P. mirabilis*. Biofilm inhibition of ≤ 99.9% (compared to the reference bead) was achieved at 1 μg/ml for 24 h. After 3 days, 5 μg/ml was needed, which decreased back to 1 μg/ml by day 5. Mixed species *P. mirabilis* exhibited the same inhibition trend when mixed with *S. epidermidis*; 0.5 μg/ml effectively inhibited both bacteria in the mixture after 5 days of exposure.

While it is evident that initial colonization of the bead (attachment of bacteria to the bead) does take place, BisEDT prevents the accumulation of biofilm. This phenomenon has been observed for other organisms (i.e. *Klebsiella pneumoniae* and *Pseudomonas aeruginosa*) when treated with varying concentrations of BisEDT (14, 35). It is possible that the initial attachment phase only consists of a monolayer on the bead (indicated by low viable cell numbers) and the BisEDT treatment suppresses cell aggregation and development into a multi-layered biofilm. However, further confirmative observations using visualizing techniques such as an electron microscope are needed to validate this hypothesis. For both organisms, the dosage necessary for the inhibition of biofilm production when existing as part of a dual species biofilm was higher than their single species counterparts. This phenomenon has been observed in other studies involving mixed biofilms produced by the same organisms and treated with gentamicin (6). The presence of a second (possibly more) organism(s) promotes shielding effects from drug activity on each other. Other investigators have reported similar findings (1, 20).

The disk diffusion susceptibility assay was done in order to measure the inhibition of growth instead of viable cell counts. Once streaked, both organisms were expected to produce biofilms on Muller-Hinton agar since the agar surface is smooth (which is a favorable condition for biofilm adherence.) This assay confirmed the observation made in the previous assay; inhibition of *P. mirabilis* requires a higher dose of the drug than *S. epidermidis*.

Polystyrene bead assay results of study #2 indicate that 5 µg/ml of BisEDT effectively eradicates both single and dual species *S. epidermidis* throughout the period viable counts were made. Single species *P. mirabilis* biofilm was effectively eradicated by 5 µg/ml except at day 1. Dual species *P. mirabilis* was eradicated at 5 µg/ml of treatment at days 1, 3 and 5. These concentrations lowered the number of viable cells by ≤ 99.9% and therefore, is the MBEC (Minimum Biofilm Eradication Concentration) according to NCCLS definitions. Eradication of a majority of bacteria is as desirable as total eradication since the immune system has been shown to play an important role in attacking remaining persistent cells (14, 19, 35).

Eradication of pre-existing biofilm requires a higher dose of BisEDT than for inhibition of biofilm formation. This is expected since the pre-formed biofilm was laid down over a 24 h. period and probably consists of multi-layers, which the drug needs to penetrate and release to the medium. Bismuth-thiols at high doses are known to penetrate biofilm layers and kill bacteria within the layer without causing them to lyse (9). It is not known if the released biofilm layers are capable of reattaching and returning to its original state. This is a possibility, especially if the BisEDT concentration in the medium sinks below the MBEC. However, the data in this study indicates effective killing up until the fifth day. A study designed to expose the biofilm longer to the drug and to investigate the stability of the drug over time may be of use.

The bead staining assay proved to be a quick and easy method of quantifying biofilm and determining treatment effects. It showed low absorbance values at higher concentrations of BisEDT for both biofilm inhibition and eradication. The uninoculated bead served as a comparison to demonstrate how well the reference bead was colonized by the bacteria. This assay also demonstrated how much biofilm remained on the bead after treatment. A disadvantage of this assay is the possibility of non-viable bacterial cells caught in the matrix contributing to higher absorbance readings. To resolve this problem, beads were sonicated for 30 sec. before staining with crystal violet (radioactive labeling was not used since both dead and live bacteria emit radioactivity.) Preliminary data showed lower absorbance levels for beads that were sonicated (compared to beads that were not sonicated), suggesting dead bacteria contributed to higher absorbance readings.

Biofilms of both organisms are more resistant to BisEDT than their planktonic counterparts as found in the MIC determination tests. *Staphylococcus epidermidis* biofilms required a five-fold increase in MIC than planktonic cells. For *P. mirabilis*, the increase was only two-fold. This finding is in harmony with published reports (25, 28, 30, 36).

An assay to determine possible bacterial antagonism between the two bacteria used was necessary to ensure biofilm inhibition/eradication observed was due to anti-biofilm properties of the drug and not due to bacteriocins produced to achieve a competitive advantage over the other species. No significant levels of antagonism were detected. This finding was confirmed with a simple antibiosis assay where each organism was centrally and counter streaked with each other. However, investigations done with other bacterial and non-bacterial species have shown significant levels of antagonism against each other when existing as a dual species biofilm (1, 20, 33).

Preparations containing bismuth compounds are widely used in medicine to treat symptoms of dyspepsia, syphilis, warts, GI (Pepto-Bismol is often prescribed to treat diarrhea and stomach ulcers) and upper respiratory infections (13). They are also used in wound dressings due to their antiseptic properties. A proposed mechanism of action of BisEDT is the interference of redox enzymes leading to depleted intracellular ATP levels. In such a state, the cell's ability to produce exopolysaccharide would be severely affected due to lack of energy (10). Additionally, recently published data indicate reduced cell-associated lipopolysaccharide, inhibition of cytotoxic protein ExoU processing, outermembrane blebbing and aggregation of cytoplasmic material (35).

No extensive work on BisEDT toxicology has been done. However, unpublished findings indicate the LD_{50} in mice for 1:1 molar ratio BisEDT is 52 mg of Bi^{3+} per kg (9). The concentration required to inhibit/eradicate biofilms are much lower as shown in this study and numerous other studies (14, 16, 35). Another advantage is that small quantities of this drug are absorbed systematically, become active locally and are degraded rapidly (9). The fact that bismuth is the least toxic of the heavy metals and that thiols are antidotes for bismuth poisoning is an added benefit.

Bismuth compounds could prove to be a solution to antibiotic resistance. MRSA (methicillin-resistant *Staphylococcus aureus*) were shown to be susceptible to subinhibitory concentrations of BisEDT (10, 11). More interestingly, BisEDT has also been shown to function synergistically with clindamycin, cefazolin, gentamicin, minocycline, vancomycin, nafcillin and gatifloxacin (11, 12). Therefore, no cross-resistance has been detected.

In conclusion, the results of this study strongly suggest the following: 1) 0.1 and 1 µg/ml of BisEDT prevented *S. epidermidis* and *P. mirabilis* biofilm, respectively. 2) 5 µg/ml eradicated 24 hour pre-formed biofilms of both bacteria. If these data can be verified in an *in vivo* model, BisEDT may prove to be useful in preventing and treating infections associated with indwelling medical devices.

TABLES

Table 1. MIC determination of BisEDT against planktonic *S. epidermidis* RP62A

Trials	BisEDT final concentrations (µg/ml)						
	0	0.05	0.1	0.5	1	5	10
1	+	+	-	-	-	-	-
2	+	+	-	-	-	-	-
3	+	+	-	-	-	-	-

Each trial was performed in duplicate. The final inoculum (approx. 5×10^5 CFU/ml) was incubated with the above concentrations of BisEDT and Mueller-Hinton broth at 37°C and shaking at 150 rpm for 24 h

(+) = visible turbid growth

(-) = no visible growth, i.e. inhibition of growth

Table 2. MIC determination of BisEDT against planktonic *Proteus mirabilis*

Trials	BisEDT final concentrations (µg/ml)						
	0	0.05	0.1	0.5	1	5	10
1	+	+	+	-	-	-	-
2	+	+	+	-	-	-	-
3	+	+	+	-	-	-	-

Each trial was performed in duplicate. The final inoculum (approx. 5×10^5 CFU/ml) was incubated with the above concentrations of BisEDT and Mueller-Hinton broth at 37°C and shaking at 150 rpm for 24 h

(+) = visible turbid growth

(-) = no visible growth, i.e. inhibition of growth

Table 3. Summary data on the inhibition of *S. epidermidis* biofilm treated with BisEDT over 5 days (polystyrene bead assay)[a]

BisEDT concentrations (μg/ml)	Mean viable cells recovered from bead CFU/ml day 1 day 3 day 5	Mean CFUlog10 per bead day 1 day 3 day 5	Mean (± SEM) inhibition of biofilm formation day 1 day 3 day 5
0 (reference)	6.6×10^6	6.82	NA
	1.1×10^6	6.05	
	9.3×10^6	6.97	
0.1	8.5×10^4	4.93	90.6 (± 5.6)
	1.7×10^5	5.24	69.0 (± 13.9)
	2.4×10^3	3.39	99.6 (± 0.3)
0.5	1.8×10^4	3.27	97.1 (± 2.3)
	5.7×10^2	2.76	97.5 (± 5.5)
	7.4×10^2	2.87	99.9 (± 0.03)

Data shown are the mean of three independent trials done in duplicate.

All tubes were inoculated with an initial inoculum of 1×10^6 - 5×10^6 CFU/ml (~ OD of 0.01)

For each trial, the cell counts recovered from the untreated bead served as the reference for comparison and determination of inhibition of the treated beads in that trial.

[a] see appendix A

Table 4. Summary data on the inhibition of dual species biofilm treated with BisEDT over 5 days and plated on MSA (polystyrene bead assay)[a]

BisEDT concentrations (µg/ml)	Mean viable cells recovered from bead CFU/ml day 1 day 3 day 5	Mean CFUlog10 per bead day 1 day 3 day 5	Mean (± SEM) inhibition of biofilm formation day 1 day 3 day 5
0 (reference)	5.8×10^8 9.3×10^7 1.1×10^7	8.77 7.97 7.06	NA
0.5	1.9×10^8 5.2×10^7 6.4×10^5	8.30 7.72 5.81	65.1 (± 3.5) 48.6 (± 11.0) 85.4 (± 7.5)
1.0	1.4×10^8 1.5×10^7 1.9×10^3	8.16 7.20 3.29	75.6 (± 6.7) 93.4 (± 6.5) 99.9 (± 0.06)
5.0	2.5×10^6 2.6×10^5 1.1×10^2	6.41 5.43 2.06	99.5 (± 0.3) 99.6 (± 0.2) 99.9 (± 0.002)

Data shown are the mean of three independent trials done in duplicate.

All tubes were inoculated with an initial inoculum of 1×10^6 - 5×10^6 CFU/ml ($\sim$ OD of 0.01)

For each trial, the cell counts recovered from the untreated bead served as the reference for comparison and determination of inhibition of the treated beads in that trial.

[a] see appendix B

Table 5. Summary data on the inhibition of *P. mirabilis* biofilm treated with BisEDT over 5 days (polystyrene bead assay)[a]

BisEDT concentrations (μg/ml)	Mean viable cells recovered from bead CFU/ml day 1 day 3 day 5	Mean CFUlog10 per bead day 1 day 3 day 5	Mean (± SEM) inhibition of biofilm formation day 1 day 3 day 5
0 (reference)	2.9×10^8 7.7×10^7 5.4×10^8	8.47 7.89 8.74	NA
0.5	2.5×10^8 3.8×10^7 2.3×10^8	8.40 7.58 8.37	21.0 (± 9.3) 52.4 (± 5.0) 54.6 (± 11.8)
1.0	7.7×10^5 2.0×10^7 1.2×10^8	5.89 7.32 8.09	99.7 (± 0.1) 60.9 (± 12.2) 75.8 (± 9.4)
5.0	1.3×10^6 1.6×10^5 1.0×10^5	6.12 5.21 5.03	99.6 (± 0.3) 99.8 (± 0.1) 99.9 (± 0.01)

Data shown are the mean of three independent trials done in duplicate.

All tubes were inoculated with an initial inoculum of 1×10^6 - 5×10^6 CFU/ml ($\sim$ OD of 0.01)

For each trial, the cell counts recovered from the untreated bead served as the reference for comparison and determination of inhibition of the treated beads in that trial.

[a] see appendix C

Table 6. Summary data on the inhibition of dual species biofilm treated with BisEDT over 5 days and plated on XLD (polystyrene bead assay)[a]

BisEDT concentrations (μg/ml)	Mean viable cells recovered from bead CFU/ml day 1 day 3 day 5	Mean CFUlog10 per bead day 1 day 3 day 5	Mean (± SEM) inhibition of biofilm formation day 1 day 3 day
0 (reference)	7.4×10^8 3.8×10^8 8.5×10^7	8.87 8.59 7.93	NA
0.5	3.5×10^8 2.2×10^8 1.0×10^7	8.55 8.35 7.04	50.5 (± 4.5) 47.9 (± 6.7) 81.9 (± 10.1)
1.0	7.0×10^7 1.2×10^8 2.1×10^5	7.85 8.11 5.33	90.8 (± 1.8) 77.3 (± 7.7) 99.2 (± 0.6)
5.0	3.6×10^7 5.3×10^6 6.6×10^2	7.56 6.73 2.79	95.9 (± 1.7) 98.8 (± 0.8) 99.9 (± 0)

Data shown are the mean of three independent trials done in duplicate.

All tubes were inoculated with an initial inoculum of 1×10^6 - 5×10^6 CFU/ml (~ OD of 0.01)

For each trial, the cell counts recovered from the untreated bead served as the reference for comparison and determination of inhibition of the treated beads in that trial.

[a] see appendix D

Table 7. Summary data on the inhibition of single species *S. epidermidis* and *P. mirabilis* biofilm treated with BisEDT for 24 h (agar diffusion susceptibility assay)[a]

BisEDT concentrations (μg per 10 μl)	Mean (SEM) diameter of the zone of inhibition of *S. epidermidis* (mm)	Mean (SEM) diameter of the zone of inhibition of *P. mirabilis* (mm)
0 (reference)	0	0
0.1	14.8 (± 0.4)	0
0.5	16.3 (± 0.8)	0
1.0	19.6 (± 0.2)	0
5.0	20.5 (± 0.3)	7.83 (± 0.1)
10.0	23.5 (± 0.2)	10 (± 0.2)

Data shown are the mean of three independent trials done in duplicate.

The reference disk (6 mm diam) contained 10 μl of distilled water and served

as the control.

[a] see appendix E

Table 8. Summary data on the inhibition of single species *S. epidermidis* and *P. mirabilis* biofilm treated with BisEDT for 24 h (spectrophotometric assay)[a]

BisEDT concentrations (μg/ml)	Mean absorbance of *S. epidermidis* biofilm on bead at 570 nm	Mean absorbance of *P. mirabilis* biofilm on bead at 570 nm
Control*	1.71	1.35
0	1.62 (-0.09)	1.67 (0.32)
0.1	1.64 (-0.07)	1.44 (0.09)
0.5	1.62 (-0.09)	1.73 (0.06)
1.0	1.35 (-0.36)	1.26 (-0.41)
5.0	1.26 (-0.45)	1.58 (0.23)
10.0	NT	1.1 (-0.25)

Data shown are the mean of three independent trials done in duplicate.

* An uninoculated bead served as the control ("blank") for each trial. The mean absorbance of the treated beads minus the blank is shown in parenthesis.

NT = Not tested

[a] see appendix F

Table 9. Summary data on the eradication of pre-formed *S. epidermidis* biofilm treated with BisEDT over 5 days (polystyrene bead assay)[a]

BisEDT concentrations (μg/ml)	Mean viable cells recovered from bead CFU/ml	Mean CFUlog10 per bead	Mean (± SEM) eradication of biofilm formation
	day 1	day 1	day 1
	day 3	day 3	day 3
	day 5	day 5	day 5
0 (reference)	6.0×10^6	6.78	NA
	1.0×10^8	8.01	
	5.1×10^6	6.71	
5.0	3.0×10^5	5.48	95.4 (± 1.8)
	4.2×10^5	5.63	98.5 (± 1.8)
	1.5×10^5	5.19	90.9 (± 8.1)
10.0	3.6×10^4	4.56	98.5 (± 1.3)
	4.3×10^3	3.64	99.7 (± 0.1)
	1.1×10^5	5.05	96.6 (± 3.2)

Data shown are the mean of three independent trials done in duplicate.

All tubes were inoculated with an initial inoculum of 1×10^6 - 5×10^6 CFU/ml (~ OD of 0.01)

For each trial, the cell counts recovered from the untreated bead served as the reference for comparison and determination of eradication of the treated beads in that trial.

[a] see appendix G

Table 10. Summary data on the eradication of pre-formed dual species biofilm treated with BisEDT over 5 days and plated on MSA (polystyrene bead assay)[a]

BisEDT concentrations (µg/ml)	Mean viable cells recovered from bead CFU/ml	Mean CFUlog10 per bead	Mean (± SEM) eradication of biofilm formation
	day 1	day 1	day 1
	day 3	day 3	day 3
	day 5	day 5	day 5
0 (reference)	3.5×10^8	8.55	NA
	9.3×10^7	7.97	
	4.6×10^7	7.67	
5.0	8.9×10^5	5.95	98.4 (± 1.1)
	3.4×10^5	5.54	97.6 (± 2.0)
	4.6×10^5	5.67	98.8 (± 0.5)
10.0	7.9×10^4	4.90	99.7 (± 0.1)
	3.8×10^3	3.59	99.9 (± 0.006)
	3.3×10^3	3.53	99.9 (± 0.002)

Data shown are the mean of three independent trials done in duplicate.

All tubes were inoculated with an initial inoculum of 1×10^6 - 5×10^6 CFU/ml (~ OD of 0.01)

For each trial, the cell counts recovered from the untreated bead served as the reference for comparison and determination of eradication of the treated beads in that trial.

[a] see appendix H

Table 11. Summary data on the eradication of pre-formed *P. mirabilis* biofilm treated with BisEDT over 5 days (polystyrene bead assay)[a]

BisEDT concentrations (µg/ml)	Mean viable cells recovered from bead CFU/ml day 1 day 3 day 5	Mean CFUlog10 per bead day 1 day 3 day 5	Mean (± SEM) eradication of biofilm formation day 1 day 3 day 5
0 (reference)	$7.2{\times}10^8$ $6.6{\times}10^8$ $4.4{\times}10^8$	8.86 8.82 8.65	NA
5.0	$6.1{\times}10^7$ $1.6{\times}10^7$ $2.0{\times}10^7$	7.79 7.23 7.31	90.2 (± 2.2) 94.4 (± 3.3) 96.1 (± 1.7)
10.0	$3.0{\times}10^4$ $3.4{\times}10^4$ $1.4{\times}10^5$	4.49 4.54 5.15	98.9 (± 0) 99.9 (± 0.02) 99.8 (± 0.04)

Data shown are the mean of three independent trials done in duplicate.

All tubes were inoculated with an initial inoculum of $1{\times}10^6$ - $5{\times}10^6$ CFU/ml (~ OD of 0.01)

For each trial, the cell counts recovered from the untreated bead served as the reference for comparison and determination of eradication of the treated beads in that trial.

[a] see appendix I

Table 12. Summary data on the eradication of pre-formed dual species biofilm treated with BisEDT over 5 days and plated on XLD (polystyrene bead assay)[a]

BisEDT concentrations (μg/ml)	Mean viable cells recovered from bead CFU/ml day 1 day 3 day 5	Mean CFUlog10 per bead day 1 day 3 day 5	Mean (± SEM) eradication of biofilm formation day 1 day 3 day 5
0 (reference)	4.6×10^8 5.6×10^8 1.2×10^8	8.67 8.75 8.11	NA
5.0	5.1×10^6 6.1×10^5 1.2×10^6	6.71 5.79 6.08	90.0 (± 0.8) 99.7 (± 0.1) 98.8 (± 0.4)
10.0	5.3×10^4 4.2×10^4 1.6×10^3	4.73 4.63 3.21	99.9 (± 0.01) 99.9 (± 0.008) 99.9 (± 0)

Data shown are the mean of three independent trials done in duplicate.

All tubes were inoculated with an initial inoculum of 1×10^6 - 5×10^6 CFU/ml (~ OD of 0.01)

For each trial, the cell counts recovered from the untreated bead served as the reference for comparison and determination of eradication of the treated beads in that trial.

[a] see appendix J

Table 13. Summary data on the eradication of single species *S. epidermidis* and *P. mirabilis* pre-formed biofilm treated with BisEDT for 24 h (spectrophotometric assay)[a]

BisEDT concentrations (μg/ml)	Mean absorbance of *S. epidermidis* biofilm on bead at 570 nm	Mean absorbance of *P. mirabilis* biofilm on bead at 570 nm
Control*	0.56	1.35
0	1.52 (0.96)	1.85 (0.5)
1.0	1.31 (0.75)	1.62 (0.27)
5.0	1.21 (0.64)	1.56 (0.21)
10.0	0.72 (0.16)	1.35 (0)

Data shown are the mean of three independent trials done in duplicate.

* An uninoculated bead served as the control ("blank") for each trial. The mean absorbance of the treated beads minus the blank is shown in parenthesis.

NT = Not tested

[a] see appendix K

Table 14. Summary data on the interactions between *S. epidermidis* and *P. mirabilis* biofilms (polystyrene bead assay)[a]

Organsim	Mean viable cells recovered from bead CFU/ml	Mean (SEM) CFUlog10 per bead
S. epidermidis (single species)	3.7×10^8	8.57 (± 0.6)
S. epidermidis (dual species)	6.3×10^8	8.8 (± 0.2)
P. mirabilis (single species)	4.6×10^8	8.67 (± 0.1)
P. mirabilis (dual species)	5.8×10^8	8.77 (± 0.1)

Data shown are the mean of three independent trials done in duplicate.

[a] see appendix L

FIGURES

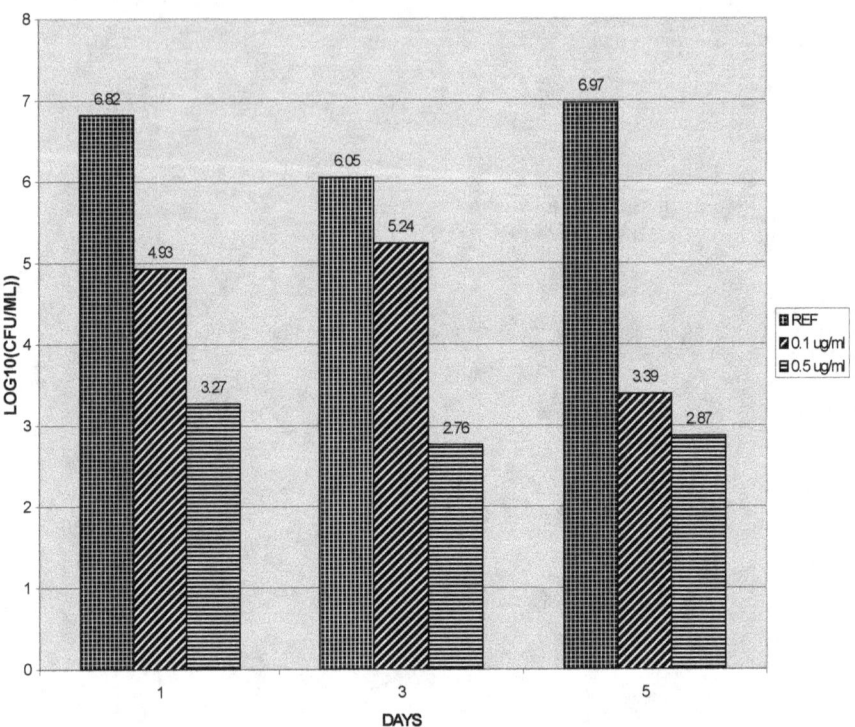

Figure 1. Inhibition of *S. epidermidis* biofilm treated with BisEDT over 5 days (polystyrene bead assay)

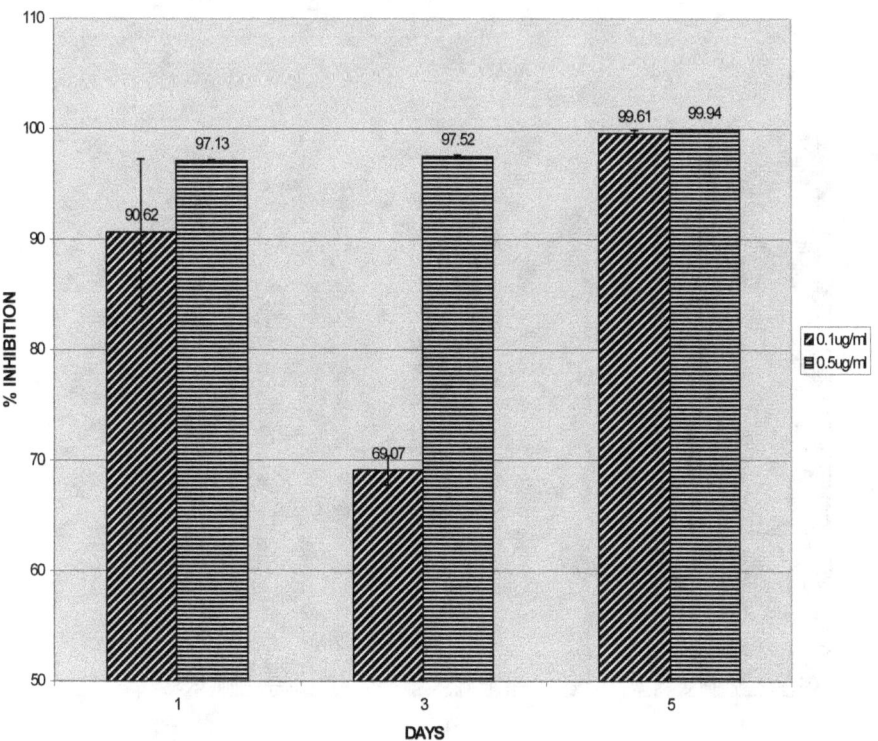

Figure 1a.The relative percent inhibition of biofilm production was determined by:

100-[(CFU of treated bead ÷ CFU of treated bead) × 100]

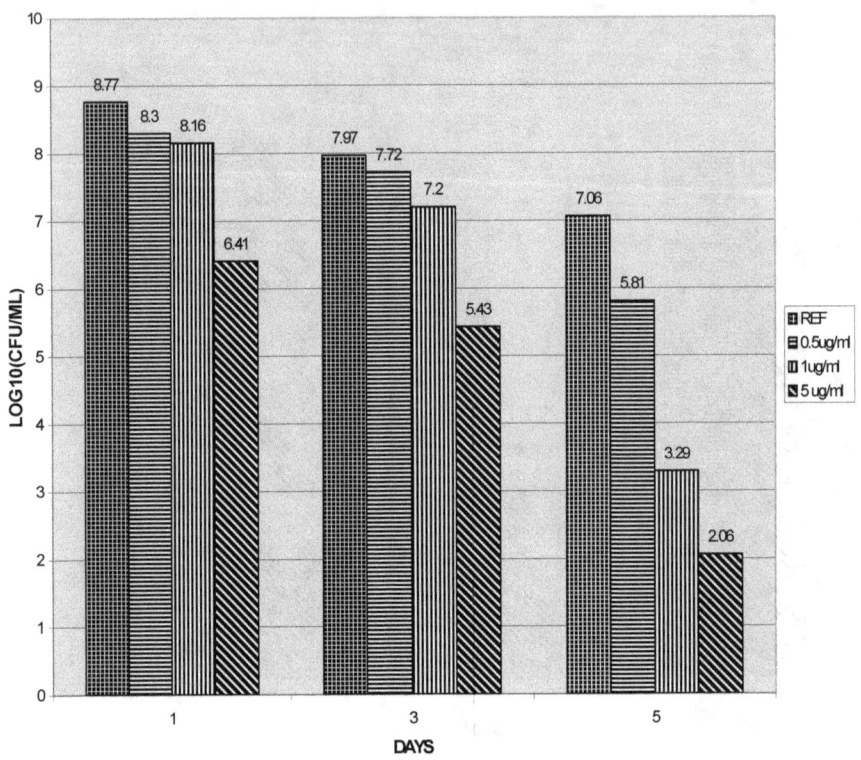

NON PREFORMED BIOFILM: MIX (MSA)

Figure 2. Inhibition of dual species biofilm treated with BisEDT over 5 days and plated on MSA (polystyrene bead assay)

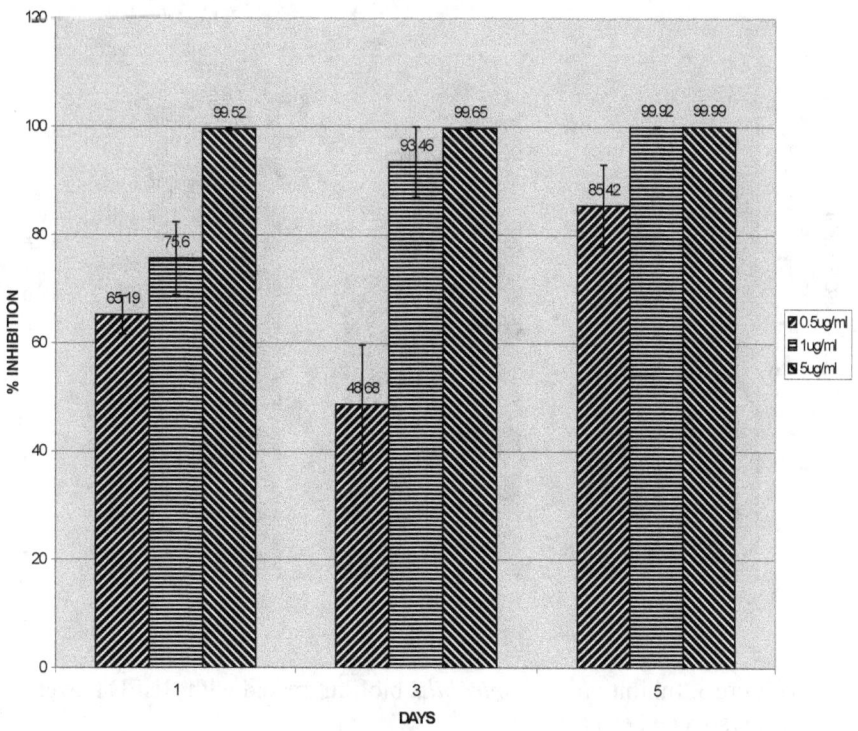

Figure 2a. The relative percent inhibition of biofilm production was determined by:

$$100-[(\text{CFU of treated bead} \div \text{CFU of treated bead}) \times 100]$$

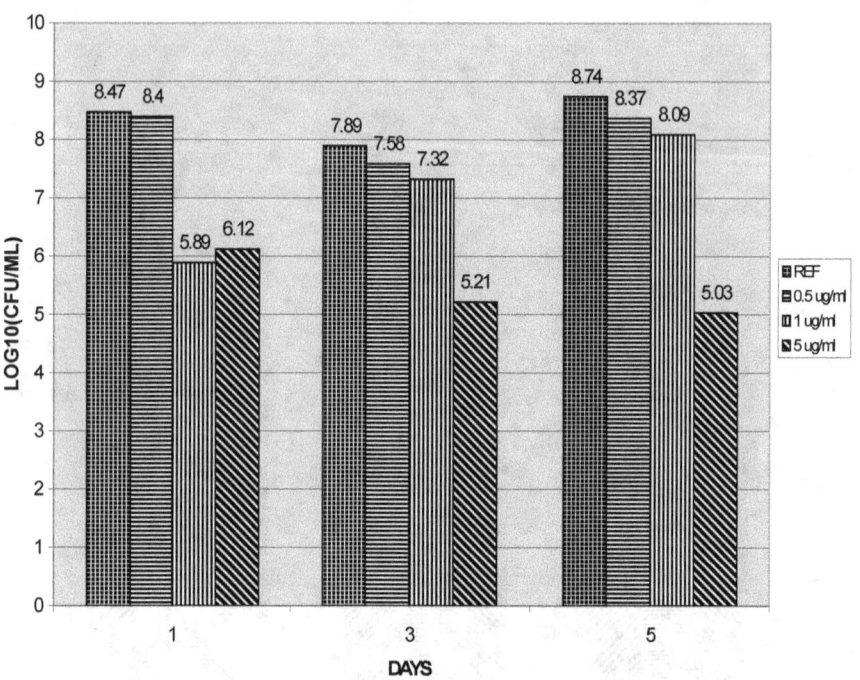

Figure 3. Inhibition of *P. mirabilis* biofilm treated with BisEDT over 5 days (polystyrene bead assay)

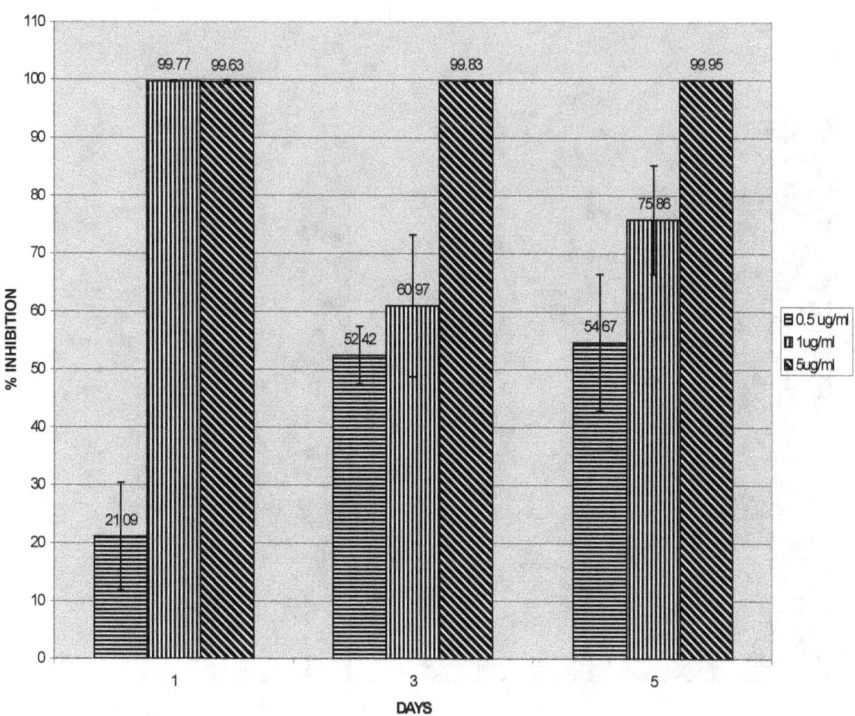

Figure 3a.The relative percent inhibition of biofilm production was determined by:

$$100-[(CFU \text{ of treated bead} \div CFU \text{ of treated bead}) \times 100]$$

NON PREFORMED BIOFILM: MIX (XLD)

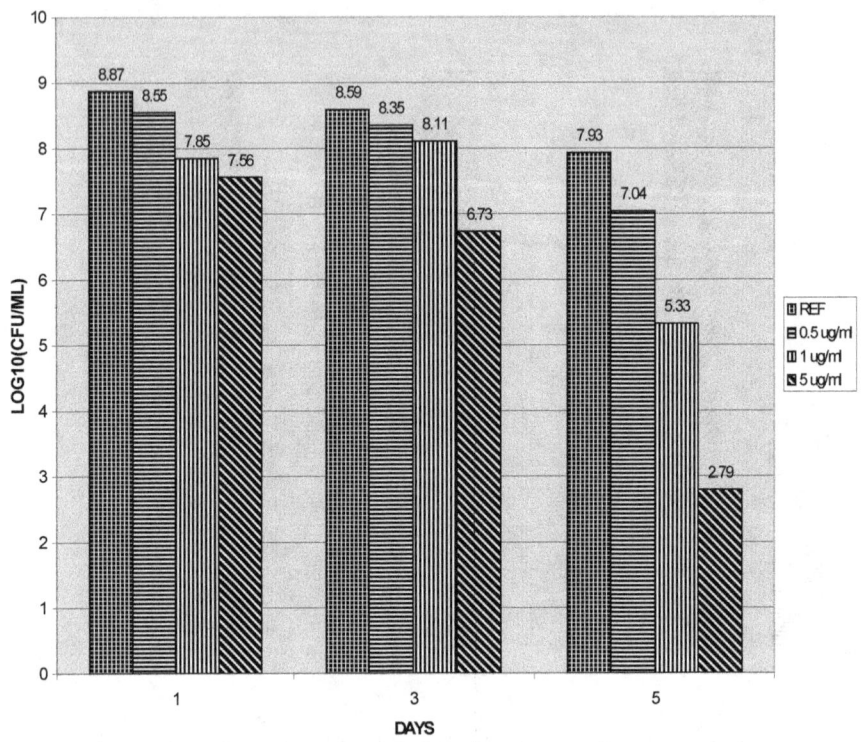

Figure 4. Inhibition of dual species biofilm treated with BisEDT over 5 days and plated on XLD (polystyrene bead assay)

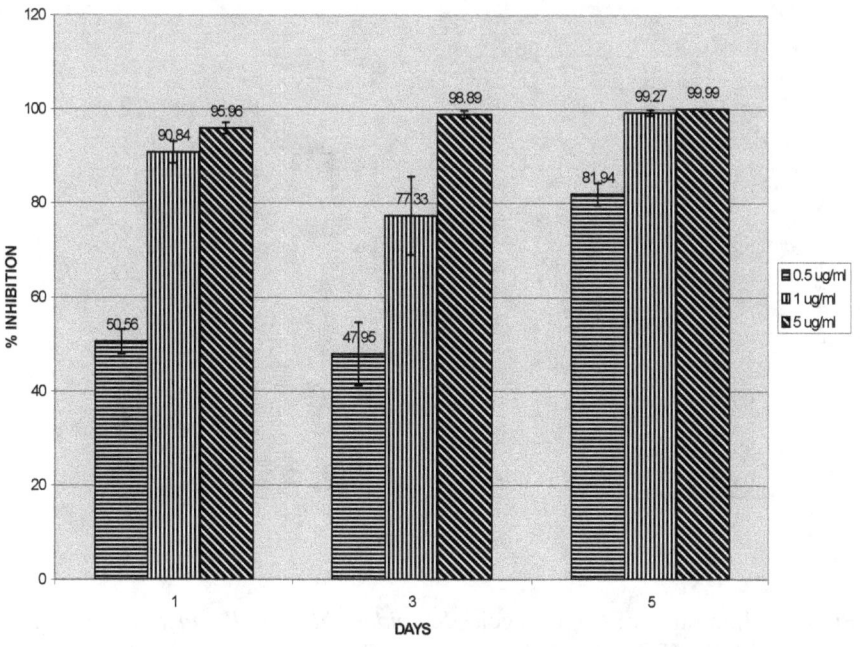

Figure 4a. The relative percent inhibition of biofilm production was determined by:

$$100-[(\text{CFU of treated bead} \div \text{CFU of treated bead}) \times 100]$$

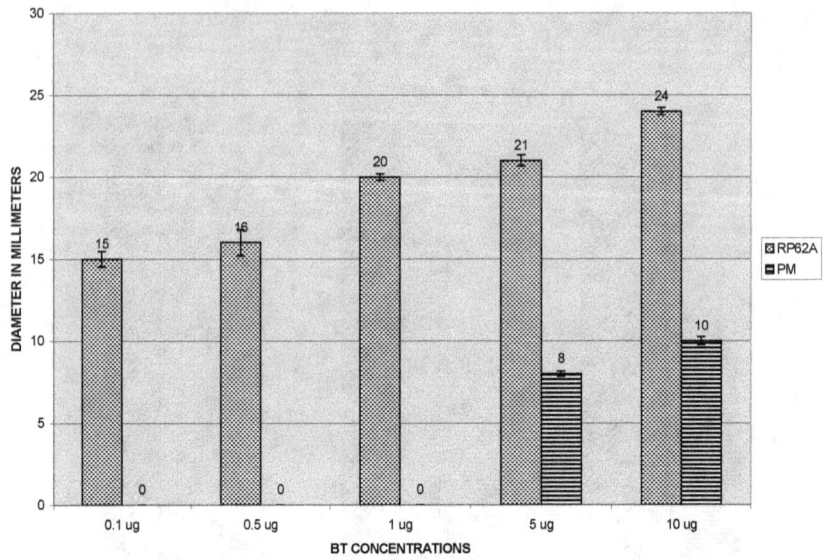

Figure 5. Inhibition of single species *S. epidermidis* and *P. mirabilis* biofilm treated with BisEDT (agar diffusion susceptibility assay)

A disk (6 mm diam.) containing distilled water served as a reference where neither organism showed inhibition to (not shown).

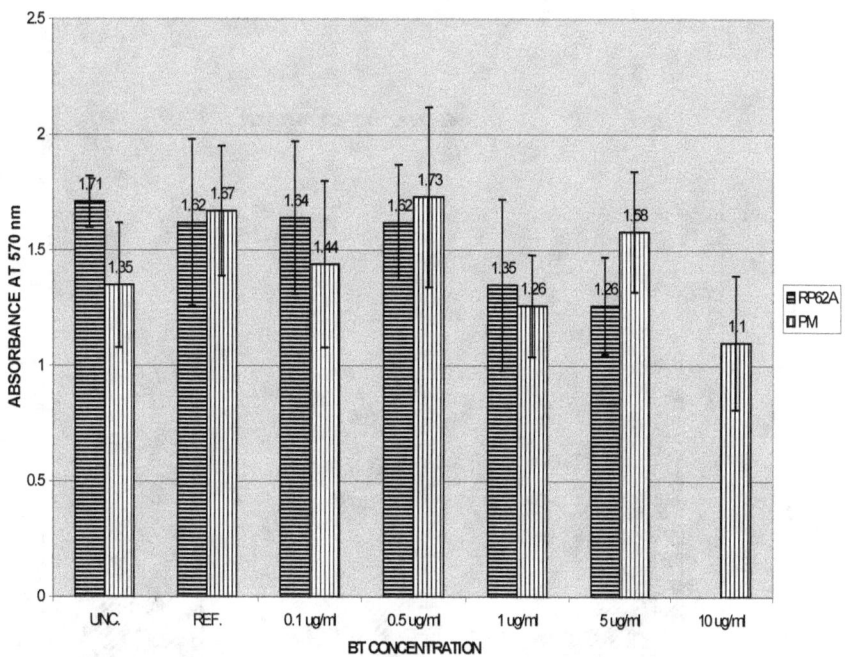

Figure 6. Inhibition of single species *S. epidermidis* and *P. mirabilis* biofilm treated with BisEDT (spectrophotometric assay)

RP62A was not treated with 10 μg/ml BisEDT.

UNC. = Uninoculated and untreated bead.

REF. = Reference bead; not treated with BisEDT.

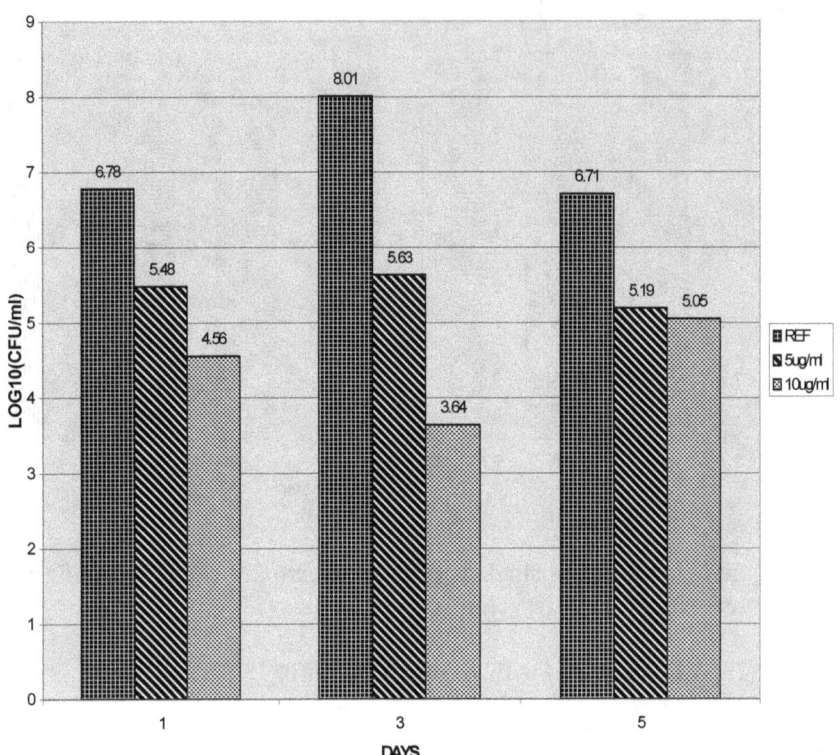

Figure 7. Eradication of pre-formed *S. epidermidis* biofilm treated with BisEDT over 5 days (polystyrene bead assay)

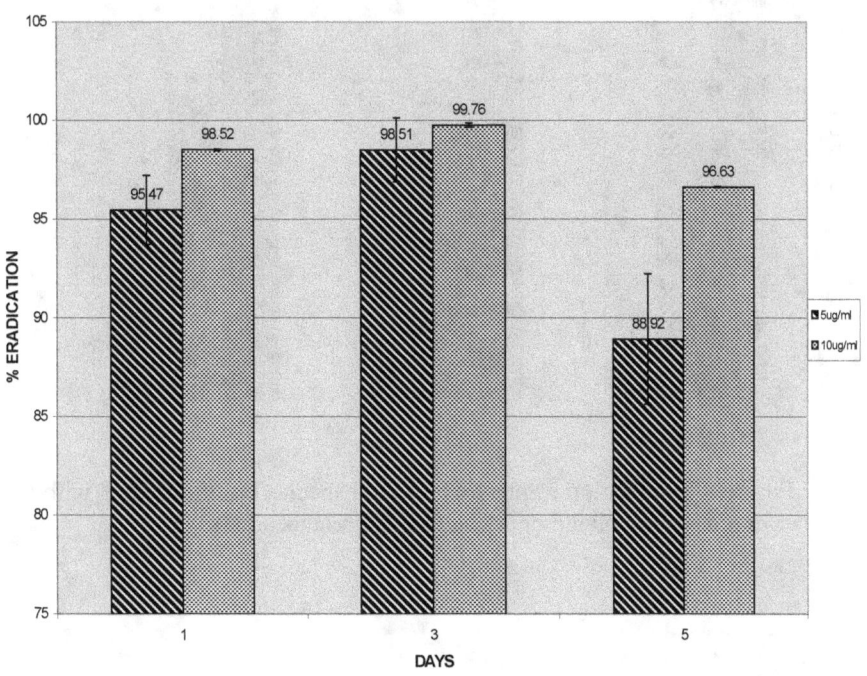

Figure 7a.The relative percent eradication of biofilm production was determined by:

$$100-[(CFU\ of\ treated\ bead \div CFU\ of\ treated\ bead) \times 100]$$

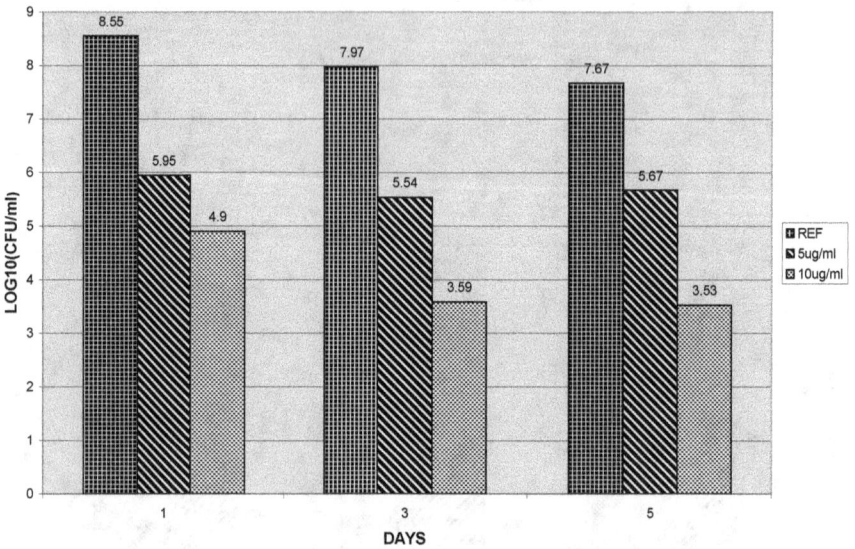

Figure 8. Eradication of pre-formed dual species biofilm treated with BisEDT over 5 days and plated on MSA (polystyrene bead assay)

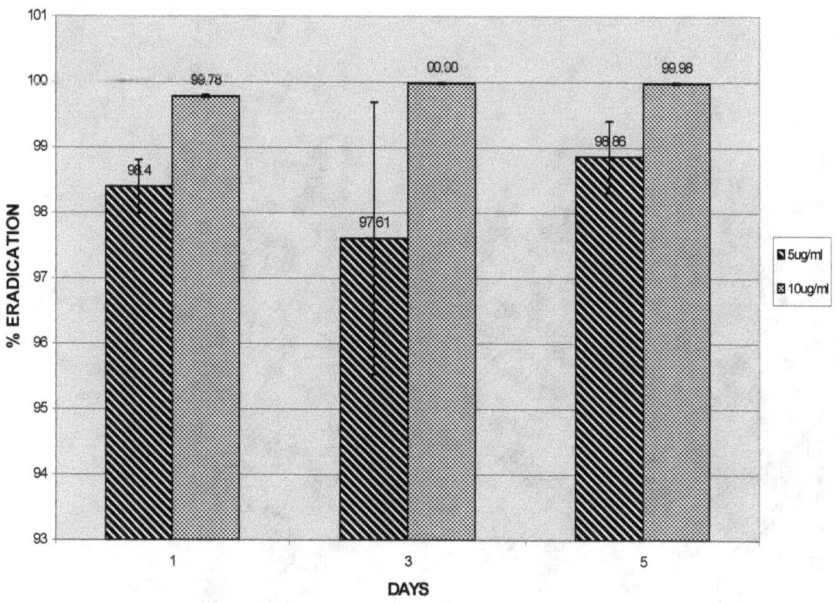

Figure 8a. The relative percent eradication of biofilm production was determined by:

$$100-[(CFU \text{ of treated bead} \div CFU \text{ of treated bead}) \times 100]$$

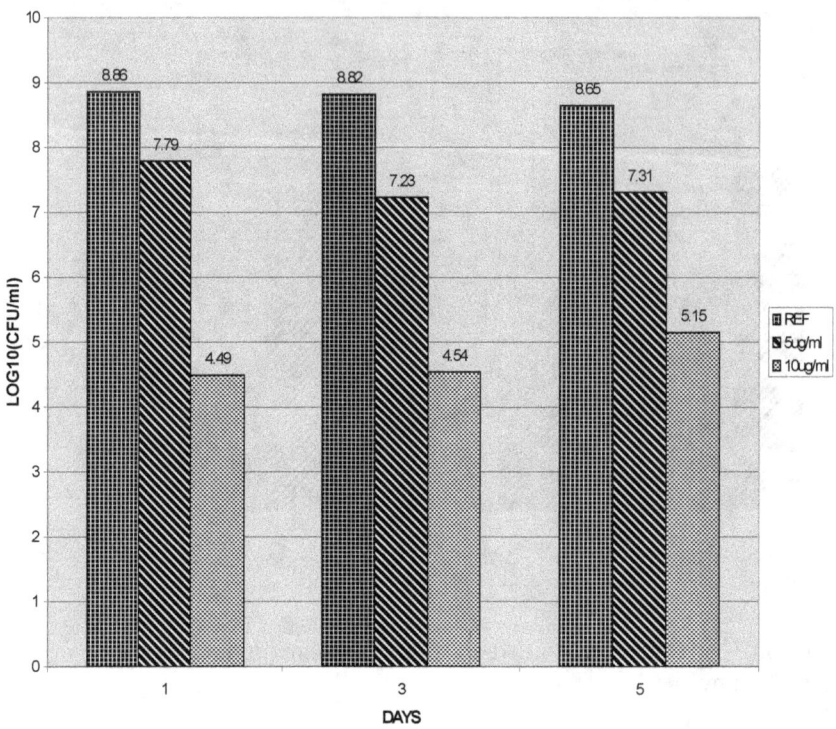

Figure 9. Eradication of pre-formed *P. mirabilis* biofilm treated with BisEDT over 5 days (polystyrene bead assay)

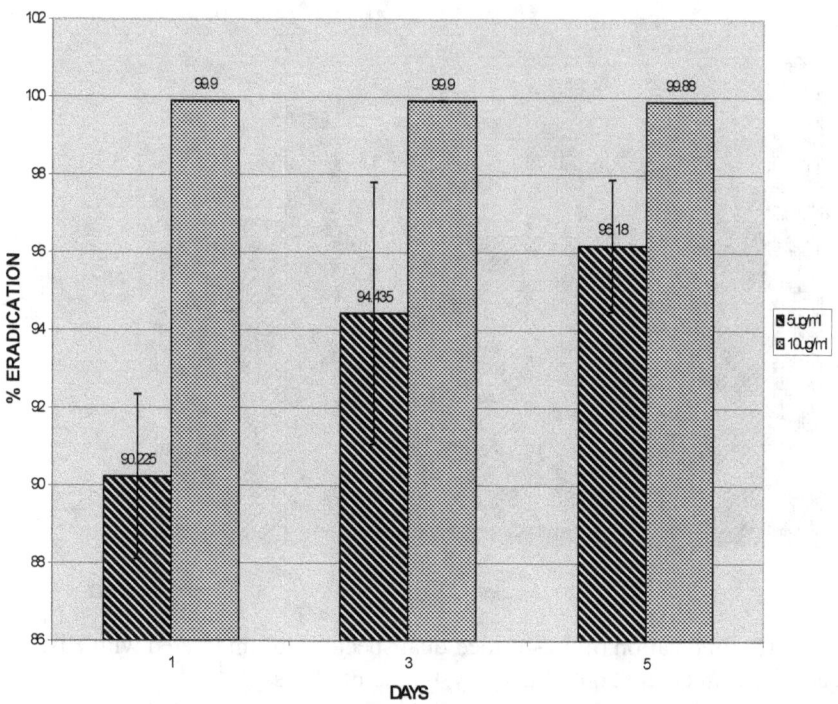

Figure 9a.The relative percent eradication of biofilm production was determined by:

100-[(CFU of treated bead ÷ CFU of treated bead) × 100]

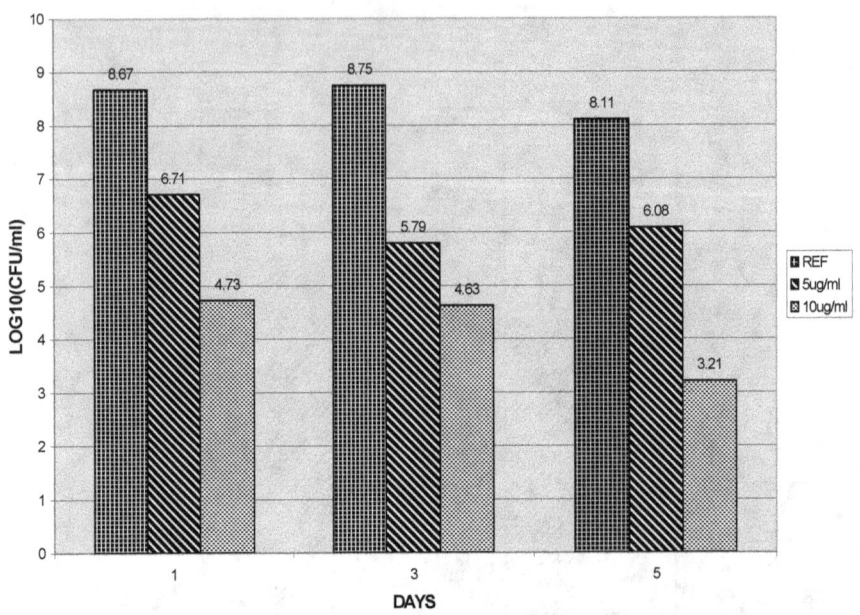

Figure 10. Eradication of pre-formed dual species biofilm treated with BisEDT over 5 days and plated on XLD (polystyrene bead assay)

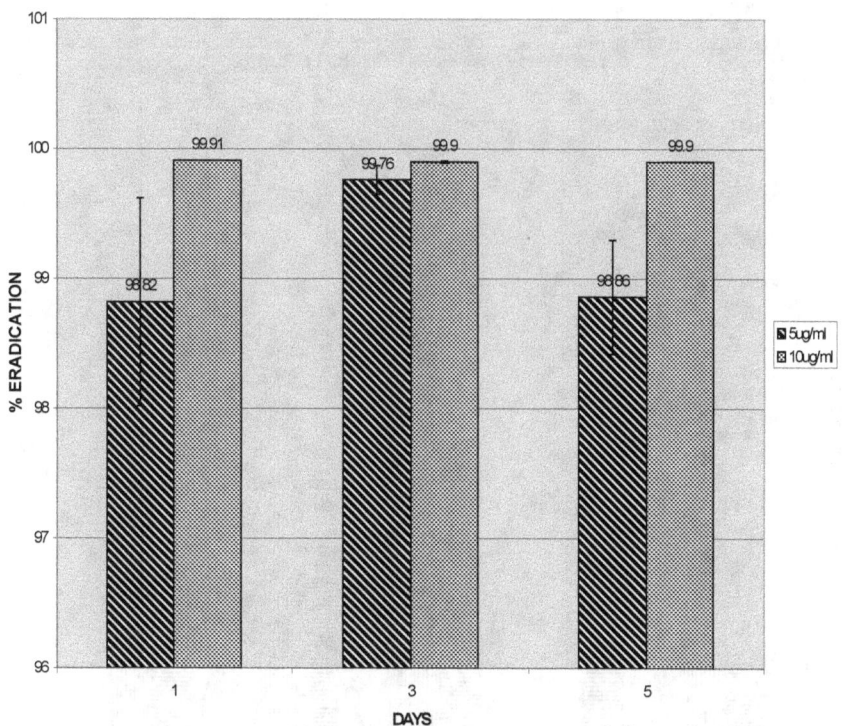

Figure 10a.The relative percent eradication of biofilm production was determined by:

100-[(CFU of treated bead ÷ CFU of treated bead) × 100]

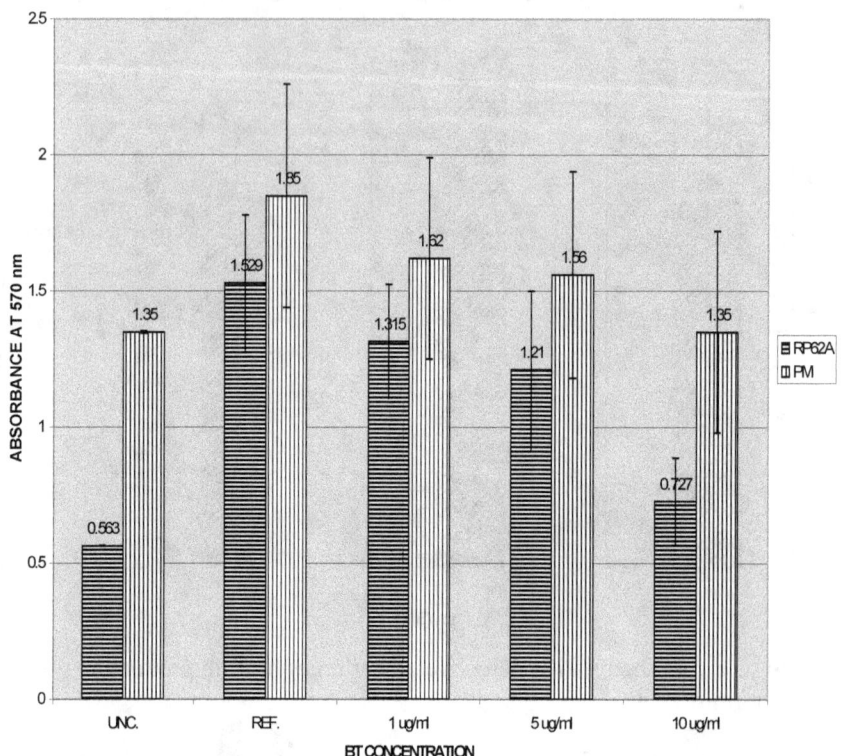

Figure 11. Eradication of single species *S. epidermidis* and *P. mirabilis* pre-
formed biofilm treated with BisEDT (spectrophotometric assay)

UNC. = Uninoculated and untreated bead.

REF. = Reference bead; not treated with BisEDT.

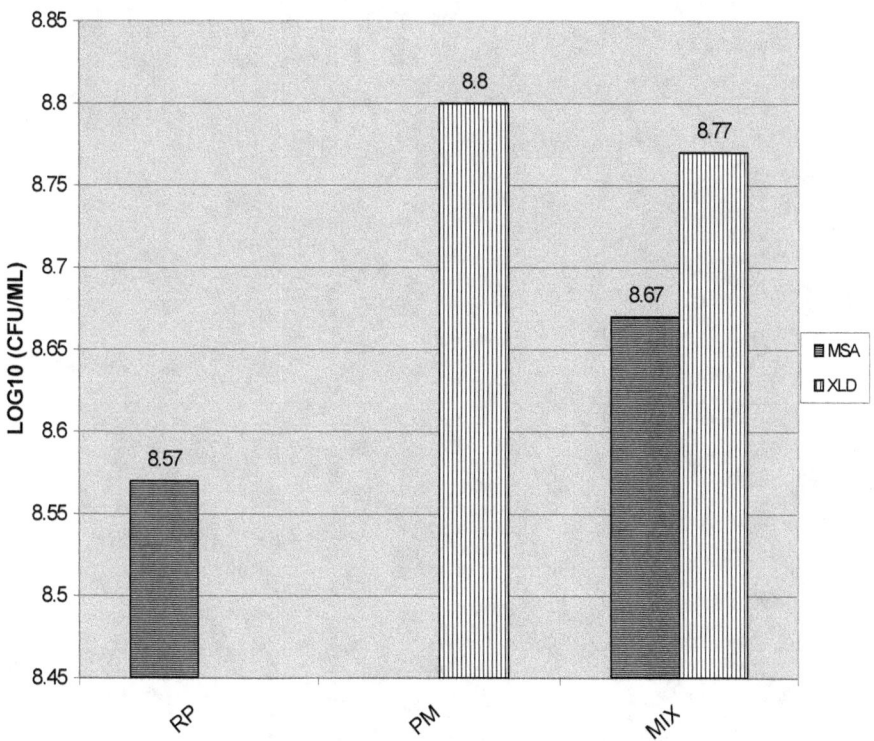

Figure 12. Interactions between *S. epidermidis* and *P. mirabilis* biofilms

MSA = Mannitol Salts Agar

XLD = Xylose Lysine Deoxycholate

APPENDICES

APPENDIX A

Inhibition of *S. epidermidis* RP62A biofilm treated with BisEDT over 5 days (polystyrene bead assay)

BisEDT Concentration (µg/ml)	Viable cells recovered after 1 day of treatment CFU/ml	% inhibition of biofilm after 1 day	Viable cells recovered after 3 days of treatment CFU/ml	% inhibition of biofilm after 3 days	Viable cells recovered after 5 days of treatment CFU/ml	% inhibition of biofilm after 5 days
Trial 1a						
REF	4.43×10^5		2.63×10^5		3.27×10^6	
0.1	9.3×10^2	99.7	1.51×10^4	4.24	3×10^3	99.9
0.5	1×10^3	99.7	2.46×10^3	99.0	7.6×10^2	99.9
Trial 1b						
REF	3.29×10^6		2.84×10^6		2.96×10^6	
0.1	1.3×10^3	99.9	1.02×10^6	63.9	4.56×10^3	99.8
0.5	1.03×10^3	99.9	1.6×10^2	99.9	1.8×10^3	99.9
Trial 2a						
REF	1.57×10^6		1.72×10^4		3.51×10^6	
0.1	8.3×10^3	99.4	1.12×10^4	34.5	1.2×10^3	99.9
0.5	1.6×10^2	99.9	0.3×10^2	99.8	7×10^2	99.9
Trial 2b						
REF	4.5×10^3		1.09×10^6		4.56×10^7	
0.1	1.03×10^3	77.1	5.26×10^3	99.5	2.3×10^3	99.9
0.5	6.6×10^2	85.3	2×10^2	99.9	2.3×10^2	99.9
Trial 3a						
REF	2.76×10^3		1.46×10^3		1.63×10^5	
0.1	8.6×10^2	68.8	1.13×10^3	22.6	3.1×10^3	98.0
0.5	0.6×10^2	97.8	2×10^2	86.3	1.6×10^2	99.9
Trial 3b						
REF	3.5×10^7		2.57×10^6		1.13×10^6	
0.1	5×10^5	98.5	1.02×10^4	99.6	8.3×10^2	99.9
0.5	83.3×10^2	99.9	4.6×10^2	99.9	8×10^2	99.9

Inhibition of dual species biofilm treated with BisEDT over 5 days and plated on MSA (polystyrene bead assay)

BisEDT Concentration (μg/ml)	Viable cells recovered after 1 day of treatment CFU/ml	% inhibition of biofilm after 1 day	Viable cells recovered after 3 days of treatment CFU/ml	% inhibition of biofilm after 3 days	Viable cells recovered after 5 days of treatment CFU/ml	% inhibition of biofilm after 5 days
Trial 1a						
REF	6.9×10^8		1.76×10^8		2.33×10^6	
0.5	1.95×10^8	71.7	1.07×10^8	39.3	1.08×10^6	53.5
1.0	2.26×10^8	67.2	1.66×10^4	99.9	9.4×10^3	99.5
5.0	1.4×10^7	97.9	1.66×10^4	99.9	NG	100
Trial 1b						
REF	6.76×10^8		6.76×10^8		2.8×10^7	
0.5	2.2×10^8	67.4	1.6×10^7	52.3	2.36×10^4	99.9
1.0	2.57×10^8	61.9	1.8×10^3	99.9	6×10^2	99.9
5.0	1.34×10^4	99.9	1.5×10^3	99.9	3×10^2	99.9
Trial 2a						
REF	8.56×10^8		1.99×10^6		1.15×10^7	
0.5	3.3×10^8	61.4	1.5×10^6	24.6	2.6×10^6	77.3
1.0	2.96×10^8	65.4	8×10^2	99.9	1.3×10^3	99.9
5.0	2.6×10^4	99.9	2.6×10^2	99.9	NG	100
Trial 2b						
REF	6.43×10^8		2.46×10^8		1.66×10^7	
0.5	2.76×10^8	57.0	1.72×10^8	30.0	2×10^4	99.8
1.0	2.09×10^7	96.7	9.6×10^7	60.9	NG	100
5.0	3.1×10^5	99.8	9.5×10^5	99.6	NG	100
Trial 3a						
REF	1.95×10^8		4.16×10^7		9.6×10^6	
0.5	8.61×10^7	55.8	2.26×10^7	45.6	2.36×10^4	99.7
1.0	6.77×10^7	65.2	1.66×10^3	99.9	5×10^2	99.9
5.0	1.35×10^6	99.3	6.8×10^5	98.3	2.6×10^2	99.9

Trial 3b						
REF	4.83×10^8		6.4×10^7		1.12×10^6	
0.5	1.08×10^8	77.6	5×10^3	99.9	2×10^5	82.1
1.0	1.46×10^7	96.9	2.12×10^4	99.9	1.71×10^2	99.9
5.0	4.1×10^3	99.9	2.5×10^3	99.9	1.35×10^2	99.9

NG = No growth observed

APPENDIX C

Inhibition of *P. mirabilis* biofilm treated with BisEDT over 5 days (polystyrene bead assay)

BisEDT Concentration (µg/ml)	Viable cells recovered after 1 day of treatment CFU/ml	% inhibition of biofilm after 1 day	Viable cells recovered after 3 days of treatment CFU/ml	% inhibition of biofilm after 3 days	Viable cells recovered after 5 days of treatment CFU/ml	% inhibition of biofilm after 5 days
Trial 1a						
REF	3.23×10^8		1.39×10^8		5.06×10^8	
0.5	1.49×10^7	53.8	4.94×10^7	64.4	2.4×10^8	52.5
1.0	8.53×10^3	99.9	1.91×10^6	98.6	3.16×10^8	37.5
5.0	1.73×10^3	99.9	1.5×10^3	99.9	1.6×10^3	99.9
Trial 1b						
REF	1.33×10^8		4.5×10^7		3.4×10^8	
0.5	9.37×10^7	29.5	2.09×10^7	53.5	1.12×10^8	66.9
1.0	1.67×10^4	99.9	1.9×10^7	57.7	2.23×10^7	93.4
5.0	3.73×10^3	99.9	1.16×10^3	99.9	3.7×10^5	99.8
Trial 2a						
REF	8.13×10^7		8.96×10^7		7.43×10^8	
0.5	7.94×10^7	2.29	5.01×10^7	44.0	8.93×10^7	87.9
1.0	1.66×10^5	99.7	2.8×10^7	68.7	2.43×10^8	66.8
5.0	1.56×10^3	99.9	1.3×10^3	99.9	1.9×10^3	99.9
Trial 2b						
REF	4.36×10^8		7.43×10^7		2.43×10^8	
0.5	5.99×10^8	?	3.39×10^7	54.3	1.18×10^8	51.4
1.0	2.14×10^2	99.9	2×10^7	73.0	7.86×10^7	67.6
5.0	3.3×10^3	99.9	6.6×10^2	99.9	2.06×10^3	99.9
Trial 3a						
REF	4.4×10^8		1.05×10^8		5.86×10^8	
0.5	3.83×10^8	12.8	7.09×10^7	32.7	5.76×10^8	1.7
1.0	4.4×10^6	99.0	4.2×10^7	60.2	9×10^6	98.4
5.0	8.83×10^3	99.9	9.83×10^5	99.0	9.83×10^5	99.9

Trial 3b						
REF	3.93×10^8		1.76×10^7		9.3×10^8	
0.5	3.66×10^8	6.87	6.1×10^6	65.3	6.26×10^3	67.4
1.0	7.8×10^4	99.9	1.63×10^7	7.38	8.16×10^7	91.2
5.0	8×10^6	97.9	6.26×10^3	99.9	2.66×10^5	99.9

APPENDIX D

Inhibition of dual species biofilm treated with BisEDT over 5 days and plated on XLD (polystyrene bead assay)

BisEDT Concentration (µg/ml)	Viable cells recovered after 1 day of treatment CFU/ml	% inhibition of biofilm after 1 day	Viable cells recovered after 3 days of treatment CFU/ml	% inhibition of biofilm after 3 days	Viable cells recovered after 5 days of treatment CFU/ml	% inhibition of biofilm after 5 days
Trial 1a						
REF	7.4×10^8		4.4×10^8		3.1×10^7	
0.5	3.8×10^8	48.6	2.53×10^8	42.5	1.76×10^7	43.2
1.0	1.04×10^8	85.9	6.6×10^7	85.0	1.3×10^6	95.0
5.0	5.1×10^7	93.1	8×10^4	99.9	6.3×10^2	99.9
Trial 1b						
REF	7.6×10^8		4.86×10^8		4.03×10^7	
0.5	3.43×10^8	54.8	3.7×10^8	23.8	1.33×10^3	99.9
1.0	5.8×10^7	92.3	3.03×10^7	93.7	1.6×10^2	99.9
5.0	4.8×10^5	99.9	2.6×10^7	94.6	0.6×10^2	99.9
Trial 2a						
REF	4.96×10^8		5.45×10^8		3.8×10^8	
0.5	2.63×10^8	46.9	2.83×10^8	48.0	3.66×10^7	90.3
1.0	5.55×10^7	88.8	2.8×10^8	48.6	8.6×10^2	99.9
5.0	1.17×10^7	97.6	6.3×10^6	98.8	1.3×10^2	99.9
Trial 2b						
REF	5.7×10^8		3.3×10^8		1.76×10^7	
0.5	3.4×10^8	40.3	1.21×10^8	63.2	2.49×10^4	99.8
1.0	5.13×10^6	99.1	1.1×10^8	66.6	1.1×10^3	99.9
5.0	4.7×10^5	99.9	2.76×10^5	99.9	6.3×10^2	99.9
Trial 3a						
REF	8.06×10^8		5.03×10^8		8.6×10^6	
0.5	4.7×10^8	41.6	3×10^8	40.3	1.45×10^4	99.8
1.0	9.1×10^7	88.7	3.6×10^6	99.2	2.3×10^2	99.9
5.0	2.86×10^7	96.4	1.36×10^4	99.9	1×10^2	99.9

Trial 3b					
REF	1.13×10^9		7.26×10^7		2.96×10^7
0.5	3.3×10^8	70.8	2.2×10^7	69.6	1.23×10^7	58.4
1.0	1.11×10^8	90.1	2.13×10^7	70.6	2.53×10^4	99.9
5.0	1.27×10^8	88.7	1.42×10^4	99.9	2.16×10^2	99.9

APPENDIX E

Inhibition of single species *S. epidermidis* and *P. mirabilis* biofilm treated with BisEDT for 24 h (agar diffusion susceptibility assay)

BisEDT concentration (µg/ml)		Diameter of zone of inhibition of *S. epidermidis RP62A* (mm)	Diameter of zone of inhibition of *P. mirabilis* (mm)
Trial 1a			
	REF	0	0
	0.1	14	0
	0.5	14	0
	1.0	19	0
	5.0	10	8
	10	23	9
Trial 1b			
	REF	0	0
	0.1	14	0
	0.5	16	0
	1.0	19	0
	5.0	20	8
	10	23	10
Trial 2a			
	REF	0	0
	0.1	14	0
	0.5	16	0
	1.0	20	0
	5.0	20	8
	10	23	10

Trial 2b

REF	0	0
0.1	15	0
0.5	16	0
1.0	20	0
5.0	20	8
10	24	10

Trial 3a

REF	0	0
0.1	15	0
0.5	16	0
1.0	20	0
5.0	21	8
10	24	10

Trial 3b

REF	0	0
0.1	17	0
0.5	20	0
1.0	20	0
5.0	22	7
10	24	11

APPENDIX F

Inhibition of single species *S. epidermidis* and *P. mirabilis* biofilm treated with BisEDT for 24 h (spectrophotometric assay)

BisEDT concentration (µg/ml)		Absorbance of *S. epidermidis RP62A* biofilm on bead at 570 nm	Absorbance of *P. mirabilis* biofilm on bead at 570 nm
Trial 1a			
	Control	1.7	1.6
	REF	2.82 (1.12)	1.9 (0.3)
	0.1	2.65 (0.95)	?
	0.5	2.39 (0.69)	1.81 (0.21)
	1.0	3.1 (1.4)	1.21 (-0.39)
	5.0	2.15 (0.45)	1.22 (-0.38)
	10	NT	2.07 (0.47)
Trial 1b			
	Control	1.74	-0.005
	REF	2.41 (0.67)	1.99 (1.99)
	0.1	2.23 (0.49)	1.48 (1.48)
	0.5	1.94 (0.2)	0.87 (0.87)
	1.0	1.32 (-0.42)	2.18 (2.18)
	5.0	1.25 (-0.49)	0.74 (0.74)
	10	NT	0.36 (0.36)
Trial 2a			
	Control	1.72	1.69
	REF	1.97 (0.25)	2.32 (0.63)
	0.1	1.71 (-0.01)	2.32 (0.63)
	0.5	1.32 (-0.4)	2.54 (0.85)
	1.0	0.81 (-0.91)	0.87 (-0.82)
	5.0	1.44 (-0.28)	2.07 (0.38)
	10	NT	1.62 (-0.07)

Trial 2b		
Control	1.72	1.7
REF	0.76 (-0.96)	2.1 (0.4)
0.1	1.75 (0.03)	2.08 (0.38)
0.5	2.12 (0.4)	0.4 (-1.3)
1.0	1.17 (-0.55)	1.67 (-0.03)
5.0	1.24 (-0.48)	1.72 (0.02)
10	NT	0.29 (-1.41)

Trial 3a		
Control	1.7	1.72
REF	0.88 (-0.82)	0.45 (-1.27)
0.1	1.14 (-0.56)	0.28 (-1.44)
0.5	1.2 (-0.5)	3.0 (1.28)
1.0	1.21 (-0.49)	0.84 (-0.88)
5.0	0.59 (-1.11)	2.54 (0.82)
10	NT	0.87 (-0.85)

Trial 3b		
Control	1.7	1.37
REF	0.88 (-0.82)	1.28 (-0.09)
0.1	0.34 (-1.36)	1.06 (-0.31)
0.5	0.75 (-0.95)	1.77 (0.4)
1.0	0.5 (-1.2)	0.82 (-0.55)
5.0	0.9 (-0.8)	1.22 (-0.15)
10	NT	1.38 (0.01)

An uninoculated bead served as the control ("blank") for each trial. The absorbance of the treated beads minus the blank is shown in parenthesis.

NT = Not tested

Eradication of pre-formed *S. epidermidis* biofilm treated with BisEDT over 5 days (polystyrene bead assay)

BisEDT Concentration (μg/ml)	Viable cells recovered after 1 day of treatment CFU/ml	% eradication of biofilm after 1 day	Viable cells recovered after 3 days of treatment CFU/ml	% eradication of biofilm after 3 days	Viable cells recovered after 5 days of treatment CFU/ml	% eradication of biofilm after 5 days
Trial 1a						
REF	1.5×10^6		1.13×10^7		2.23×10^7	
5	1.7×10^3	99.8	4.86×10^3	99.9	2.47×10^4	99.8
10	4.6×10^2	99.9	1×10^2	99.9	1.26×10^3	99.9
Trial 1b						
REF	1.6×10^4		2.5×10^6		3.4×10^6	
5	1.93×10^3	87.9	2.06×10^5	91.7	5.63×10^5	83.4
10	3.44×10^4	?	1.51×10^4	99.3	6.7×10^5	80.2
Trial 2a						
REF	2.57×10^6		1.48×10^6		6.93×10^5	
5	1.73×10^5	93.2	2×10^2	99.9	3.43×10^5	50.5
10	1.76×10^5	93.1	7.4×10^3	99.5	1.4×10^3	99.7
Trial 2b						
REF	6.1×10^6		1.43×10^8		1.98×10^6	
5	1.2×10^5	98.0	4×10^5	99.7	8.6×10^2	99.9
10	5×10^3	99.9	2.33×10^3	99.9	NG	100
Trial 3a						
REF	2.63×10^7		4.4×10^8		2.2×10^6	
5	1.54×10^6	94.1	6.7×10^3	99.9	3.26×10^3	99.8
10	1.8×10^3	99.9	3×10^2	99.9	1.76×10^3	99.9
Trial 3b						
REF	4×10^5		2.1×10^7		5×10^5	
5	1.46×10^3	99.6	1.96×10^6	90.6	4×10^2	99.9
10	9.3×10^2	99.7	1.42×10^3	99.9	0.3×10^2	99.9

Eradication of pre-formed dual species biofilm treated with BisEDT over 5 days and plated on MSA (polystyrene bead assay)

BisEDT Concentration (µg/ml)	Viable cells recovered after 1 day of treatment CFU/ml	% eradication of biofilm after 1 day	Viable cells recovered after 3 days of treatment CFU/ml	% eradication of biofilm after 3 days	Viable cells recovered after 5 days of treatment CFU/ml	% eradication of biofilm after 5 days
Trial 1a						
REF	1.04×10^7		1.09×10^6		4.2×10^7	
5	7.3×10^5	92.9	1.4×10^5	87.2	1.03×10^6	97.5
10	1.1×10^5	98.9	5.3×10^2	99.9	3.3×10^2	99.9
Trial 1b						
REF	1.64×10^8		5×10^7		6.9×10^7	
5	3.6×10^6	97.8	1.66×10^4	99.9	2.13×10^5	99.6
10	3.26×10^5	99.8	3×10^2	99.9	4.83×10^3	99.9
Trial 2a						
REF	8.56×10^8		9.43×10^7		3.7×10^7	
5	6.46×10^5	99.9	3.6×10^5	99.6	1.183×10^6	96.8
10	1.75×10^4	99.9	1.51×10^4	99.9	5.3×10^2	99.9
Trial 2b						
REF	3.5×10^8		2.02×10^8		2.6×10^7	
5	2.44×10^4	99.9	1.93×10^3	99.9	8.3×10^3	99.9
10	4.3×10^3	99.9	2.03×10^3	99.9	2.73×10^3	99.9
Trial 3a						
REF	8.56×10^7		7.8×10^7		5.3×10^7	
5	1.93×10^5	99.7	1.42×10^4	99.9	1.886×10^4	99.9
10	2.23×10^3	99.9	2.26×10^3	99.9	8.4×10^3	99.9

Trial 3b						
REF	6.96×10^8		1.46×10^8		5.46×10^7	
5	2.7×10^5	99.9	1.58×10^6	98.9	4.03×10^5	99.2
10	2.55×10^4	99.9	3.5×10^3	99.9	3.66×10^3	99.9

APPENDIX I

Eradication of pre-formed *P. mirabilis* biofilm treated with BisEDT over 5 days (polystyrene bead assay)

BisEDT Concentration (µg/ml)	Viable cells recovered after 1 day of treatment CFU/ml	% eradication of biofilm after 1 day	Viable cells recovered after 3 days of treatment CFU/ml	% eradication of biofilm after 3 days	Viable cells recovered after 5 days of treatment CFU/ml	% eradication of biofilm after 5 days
Trial 1a						
REF	8.5×10^8		1.28×10^8		1.10×10^8	
5	1.03×10^8	87.8	2.26×10^7	82.3	2.37×10^6	97.8
10	1.81×10^4	99.9	2.23×10^3	99.9	6.66×10^3	99.9
Trial 1b						
REF	6.1×10^8		1.92×10^8		6×10^8	
5	3.23×10^7	94.7	1.6×10^6	99.1	5.93×10^7	90.1
10	1.78×10^4	99.9	1.43×10^3	99.9	6.1×10^3	99.9
Trial 2a						
REF	8.53×10^8		4.66×10^8		5.5×10^8	
5	1.24×10^8	85.3	6.8×10^7	85.4	4.56×10^7	91.7
10	1.33×10^5	99.9	2.03×10^5	99.9	8.4×10^5	99.8
Trial 2b						
REF	1.68×10^8		7.66×10^8		1.62×10^8	
5	2.73×10^7	83.7	9.55×10^6	99.7	9.16×10^5	99.4
10	1.34×10^4	99.9	1.06×10^3	99.9	2.3×10^2	99.9
Trial 3a						
REF	6.6×10^8		1.25×10^9		5×10^8	
5	5.06×10^7	92.3	1.02×10^4	99.9	1.44×10^6	99.7
10	6.23×10^3	99.9	1.53×10^3	99.9	1×10^2	99.9
Trial 3b						
REF	1.22×10^9		1.16×10^9		7.83×10^8	
5	3.16×10^7	97.4	6.36×10^5	99.9	1.3×10^7	98.3
10	8.3×10^2	99.9	3.3×10^2	99.9	2×10^2	99.9

APPENDIX J

Eradication of pre-formed dual species biofilm treated with BisEDT over 5 days and plated on XLD (polystyrene bead assay)

BisEDT Concentration (μg/ml)	Viable cells recovered after 1 day of treatment CFU/ml	% eradication of biofilm after 1 day	Viable cells recovered after 3 days of treatment CFU/ml	% eradication of biofilm after 3 days	Viable cells recovered after 5 days of treatment CFU/ml	% eradication of biofilm after 5 days
Trial 1a						
REF	4.9×10^8		5.63×10^8		1.71×10^8	
5	2.66×10^7	94.5	1.38×10^6	99.7	1.65×10^6	99.0
10	7.73×10^3	99.9	5.6×10^2	99.9	1×10^2	99.9
Trial 1b						
REF	2.12×10^8		2.16×10^8		1.16×10^8	
5	1.71×10^6	99.1	1.83×10^5	99.9	1.61×10^6	98.6
10	2.93×10^5	99.8	9×10^2	99.9	8.03×10^3	99.9
Trial 2a						
REF	8.53×10^8		5.63×10^8		8×10^7	
5	1.22×10^6	99.8	1.05×10^6	99.8	2.50×10^6	96.8
10	1.27×10^4	99.9	2.56×10^5	99.9	0.3×10^2	99.9
Trial 2b						
REF	4.56×10^8		1.26×10^9		1.97×10^8	
5	3.1×10^5	99.9	2.43×10^3	99.9	2.53×10^5	99.8
10	3×10^2	99.9	3.3×10^3	99.9	1×10^2	99.9
Trial 3a						
REF	9.2×10^7		6.43×10^8		1.31×10^8	
5	4×10^5	99.5	1.3×10^4	99.9	7.56×10^5	99.4
10	1.03×10^3	99.9	8.6×10^2	99.9	1.5×10^3	99.9
Trial 3b						
REF	7.26×10^8		1.34×10^8		7.66×10^7	
5	8.83×10^5	99.8	1.13×10^6	99.1	4.76×10^5	99.3
10	9.2×10^3	99.9	1×10^2	99.9	1.6×10^2	99.9

Eradication of single species *S. epidermidis* and *P. mirabilis* biofilm treated with BisEDT for 24 h (spectrophotometric assay)

BisEDT concentration (µg/ml)	Absorbance of *S. epidermidis RP62A* biofilm on bead at 570 nm	Absorbance of *P. mirabilis* biofilm on bead at 570 nm
Trial 1a		
Control	0.56	1.37
REF	2.41(1.85)	3.0 (1.63)
1.0	1.67 (1.11)	2.59 (1.22)
5.0	0.9 (0.34)	1.61 (0.24)
10	0.32 (-0.24)	1.52 (0.15)
Trial 1b		
Control	0.55	1.35
REF	1.01 (0.46)	3.0 (1.65)
1.0	0.47 (-0.08)	2.39 (1.04)
5.0	1.05 (0.5)	2.98 (1.63)
10	0.19 (-0.36)	3.0 (1.65)
Trial 2a		
Control	0.56	1.37
REF	1.98 (1.42)	1.96 (0.59)
1.0	1.71 (1.15)	1.73 (0.36)
5.0	2.6 (2.04)	2.16 (0.79)
10	0.99 (0.43)	1.47 (0.1)

Trial 2b		
Control	0.56	1.35
REF	1.35 (0.79)	1.07 (-0.28)
1.0	1.84 (1.28)	2.02 (0.67)
5.0	1.33 (0.77)	1.52 (0.17)
10	1.28 (0.72)	0.53 (-0.82)

Trial 3a		
Control	0.56	1.36
REF	1.67 (1.11)	1.66 (0.3)
1.0	1.11 (0.55)	0.76 (-0.6)
5.0	0.71 (0.15)	0.7 (-0.66)
10	0.7 (0.14)	1.12 (-0.24)

Trial 3b		
Control	0.56	1.34
REF	0.73 (0.17)	0.44 (-0.9)
1.0	1.06 (0.5)	0.26 (-1.08)
5.0	0.66 (0.1)	0.38 (-0.96)
10	0.85 (0.29)	0.45 (-0.89)

An uninoculated bead served as the control ("blank") for each trial. The absorbance of the treated beads minus the blank is shown in parenthesis.

NT = Not tested

Interactions between *S. epidermidis* and *P. mirabilis* biofilms polystyrene bead assay)

Organism	Viable counts recovered from bead after 24 hours CFU/ml
Trial 1a	
S. epidermidis (single species)	1.43×10^6
S. epidermidis (dual species)	4.1×10^7
P. mirabils (single species)	8.93×10^7
P. mirabilis (dual species)	1.21×10^8
Trial 1b	
S. epidermidis (single species)	2×10^6
S. epidermidis (dual species)	1.46×10^7
P. mirabils (single species)	1.58×10^8
P. mirabilis (dual species)	1.13×10^8
Trial 2a	
S. epidermidis (single species)	6.23×10^8
S. epidermidis (dual species)	8.86×10^8
P. mirabils (single species)	1.033×10^9
P. mirabilis (dual species)	1.006×10^9

Trial 2b	
S. epidermidis (single species)	1.73×10^5
S. epidermidis (dual species)	4.9×10^7
P. mirabils (single species)	1.24×10^8
P. mirabilis (dual species)	1.18×10^8
Trial 3a	
S. epidermidis (single species)	5.66×10^8
S. epidermidis (dual species)	7.76×10^8
P. mirabils (single species)	8.6×10^8
P. mirabilis (dual species)	1.07×10^9
Trial 3b	
S. epidermidis (single species)	5.46×10^8
S. epidermidis (dual species)	8.43×10^8
P. mirabils (single species)	1.19×10^9
P. mirabilis (dual species)	9.76×10^8

Staphylococcus epidermidis was plated on MSA and *P. mirabilis* was plated on XLD.

BIBLIOGRAPHY

(1) Adam, B., G. S. Baillie and L. J. Douglas. (2002) Mixed Species Biofilms of *Candida albicans* and *Staphylococcus epidermidis*. *Journal of Medical Microbiology* 51: 344-9

(2) Bio Med center: Catheter problems (http://www.bio-medcen.com/catheter_problems.htm) (05/01)

(3) Bomchil, N., P. Watnick and R. Kolter. (2003) Identification and Characterization of a *Vibrio cholerae* Gene, *mbaA*, Involved in Maintenance of Biofilm Architecture. *Journal of Bacteriology* 185: 1384-90

(4) Brooks, T. and C. W. Keevii. (1997) A simple artificial urine for the growth of urinary pathogens. *Letters in Applied Microbiology* 24: 203-6

(5) Christensen, G. D., W. A. Simpson, J. J. Younger, L. M. Badddour, F. F. Barrett, D. M. Melton and E. H. Beachey. (1985) Adherence of Coagulase-Negative Staphylococci to Plastic Tissue Culture Plates: a Quantitative Model for the Adherence of Staphylococci to Medical Devices. *Journal of Clinical Microbiology* 22: 996-1006

(6) Chang, C. C. and K. Merritt. (1991) Effect of *Staphylococcus epidermidis* on Adherence of *Pseudomonas aeruginosa* and *Proteus mirabilis* to Polymethyl Methacrylate (PMMA) and Gentamicin-containing PMMA. *Journal of Orthopaedic Research* 9: 284-8

(7) Costerton, J.W., Stewart P. S., Greenberg E. P. (1999) Bacterial Biofilms: A Common Cause of Persistent Infections. *Science* 284: 1318-22

(8) **De Silva, G. D. I., M. Kanzanou, A. Justice, R. C. Massey, A. R. Wilkinson, N. P. J Day and S. J. Peacock.** (2002) The *ica* Operon and Biofilm production in Coagulase-Negative Staphylococci Associated with Carriage and Disease in a Neonatal Intensive Care Unit. *Journal of Clinical Microbiology* 40: 382-388

(9) **Domenico, P.** Bismuth-thiols (BTs) as Anti-Biofilm Agents (personal communication)

(10) **Domenico, P., L. Baldassarri, P. E. Schoch, K. Kaehler, M. Sasatsu and B. A. Cunha.** (2001) Activities of Bismuth Thiols against *Staphylococci* and *Staphylococcus* Biofilms. *Antimicrobial Agents and Chemotherapy* 45: 1417-1421

(11) **Domenico P., J. A. Kazzaz and J. M. Davis.** Synergy of Nafcillin or Gentamicin with Bismuth-Thiols (BTs) Against resistant *Staphylococcus aureus.* (personal commuication)

(12) **Domenico P., J. A. Kazzaz and J. M. Davis.** Bismuth-Thiols Prevent Biofilm Formation and Enhance Sensitivity of *S. epidermidis* to Antibiotics. (personal commuication)

(13) **Domenico, P., R. J. Salo, S. G. Novick, P. E. Schoch, K. V. Horn and B. A. Cunha** (1997) Enhancement of Bismuth Antibacterial Activity with Lipophilic Thiol Chelators. *Antimicrobial Agents and Chemotherapy* 41: 1697-1703

(14) **Domenico, P., J. M. Thomas, S. Merino, X. Rubires and B. A. Cunha.** (1999) Surface Antigen Exposure by Bismuth Dimercaprol Suppression of *Klebsiella pneumoniae* Capsular Polysaccharide. *Infection and Immunity* 67: 664-69

(15) Donlan, R. (2002) Biofilms: Microbial Life on Surfaces. Emerging Infectious Disease. 18: (http://www.cdc.gov) (06/01)

(16) Dunne, M. (2002) Bacterial Adhesion: Seen Any Good Biofilms Lately? *American Society for Microbiology* 15: 155-166

(17) Eiff, C. V., G. P. Peters. (1999). New aspects in the Molecular Basis of Polymer-Associated Infections due to Staphylococci. *European Journal of Clinical Microbiology and Infectious Disease* 18: 842-846

(18) Huang, C. T. and P. Stewart. (1999) Reduction of Polysaccharide production in *Pseudomonas aeruginosa* Biofilms by Bismuth Dimercaprol (BisBAL) Treatment. *Journal of Antimicrobial Chemotherapy* 44: 601-5

(19) Lewis, K. (2001) Mini review-Riddle of Biofilm Resistance. *Antimicrobial Agents and Chemotherapy* 45: 999-1007

(20) Lindsay, D., V. S. Brozel, J. F. Mostert and A. von Holy. (2002) Differential Efficacy of a Chlorine Dioxide-Containing Sanitizer Against Single Species and Binary Biofilms of a Dairy-Associated *Bacillus cereus* and a *Pseudomonas fluorescens* isolate. *Journal of Applied Microbiology* 92: 352-61

(21) Mack, D., N. Siemssen and R. Laufs. (1992) Parallel Induction by Glucose of Adherence and a Polysaccharide Antigen Specific for Plastic-Adherent *Staphylococcus epidermidis*: Evidence for Functional Relation to Intracellular Adhesion. *Infection and Immunity* 60: 2048-57.

(22) O'gara, J. P. and H. Humphreys. (2001) *Staphylococcus epidermidis* Biofilms: Importance and Implications. *Journal of Medical Microbiology* 50: 582-7

(23) Otto, M., H. Echner, W. Voelter and F. Gotz. (2001) Pheromone Cross-Inhibition between *Satphylococcus aureus* and *Staphylococcus epidermidis*. *American Society for Microbiology News* 69: 1957-60

(24) Polonio, R. E. (2000) Sodium Salicylate: A Potential Adjuvant to Vancomycin in Device-Related Infections Involving Biofilms of *Staphylococcus epidermidis.* Master's thesis, University of Rhode Island

(25) Polonio, R. E., L. A. Mermel, G. E. Paquette and J. F. Sperry. (2000) Eradication of Biofilm-Forming *Staphylococcus epidermidis* (RP62A) by a Combination of Sodium Salicylate and Vancomycin. *Antimicrobial Agents and Chemotherapy* 45: 3262-3266

(26) Pittsburg Regional Healthcare Initiative Infection Control Programs (http://www.prhi.org) (04/02)

(27) Saint, S. Prevention of Nosocomial Urinary Tract Infections. (http://www.ahcpr.gov/clinic/ptsafety/chap15a.htm) (06/02)

(28) Schwank, S., Z. Rajacic, W. Zimmerli and J. Blaser. (1998) Impact of Bacterial Biofilm Formation on In Vitro and In Vivo Activities of Antibiotics. *Antimicrobial Agents and Chemotherapy* 42: 895-8

(29) Sieradzki, K., P. Villari and A. Tomasz. (1998) Decreased Susceptibilities to Teicoplanin and Vancomycin among Coagulase-Negative Methicillin-Resistant Clinical Isolates of Staphylococci. *Antimicrobial Agents and Chemotherapy* 42: 100-7

(30) **Stickler, D., J. Dolman, S. Rolfe and J . Chawla.** (1991) Activity of Some Antiseptics against Urinary Tract Pathogens Growing as Biofilms on Silicone Surfaces. *European Journal of Clinical Microbiology and Infectious Disease* 10: 410-15

(31) **Stickler, D., N. Morris, M. C. Moreno and N. Subbuba.** (1998) Studies on the Formation of Crsytalline Bacterial Biofilms on Urethral Catheters. *European Journal of Clinical Microbiology and Infectious Disease* 17: 649-52

(32) **Tait, K. and I. W Sutherland.** (2002) Antagonistic Interactions Amongst Bacteriocin-producing Enteric Bacteria in Dual Species Biofilms. *Journal of Applied Microbiology* 93: 345-54

(33) **Tenover, F.C., M. V. Lancaster, B. C. Hill, C. D. Steward, S. A. Stocker, G. A. Hancock, C. M. O'Hara, N. C. Clark and K. Hiramatsu.** (1998) Characterization of Staphylococci with Reduced Susceptibilities to Vancomycin and Other Glycopeptides. *Journal of Clinical Microbiology* 36: 1020-27

(34) **Woods, G. L and J. A. Washignton.** 1995 Antibacterial Susceptibility Tests: Dilution and Disk Diffusion Methods p. 1327-41. *In* **P. R. Murray. E. J. Baron, M. A. Pealler, F. C. Tenover and R. H. Yolken.** Manual of Clinical Microbiology (6th ed.) American Society for Microbiology Press, Washington D.C.

(35) **Wu, C.L., P. Domenico, D. J. Hassett, T. J. Beveridge, A. R. Hauser and J. A. Kazzaz.** (2002) Subinhibitory Bismuth-Thiols Reduce Virulence of *Pseudomonas aeruginosa. American Journal of Respiratory Cell and Molecular Biology* 26: 731-8

(36) Zheng, Z and P. S. Stewart. (2002) Penetration of Rifampin through *Staphylococcus epidermidis* biofilms. *Antimicrobial Agents and Chemotherapy* 46: 900-3

www.ingramcontent.com/pod-product-compliance
Lightning Source LLC
Chambersburg PA
CBHW081141170526
45165CB00008B/2750